MAKING TIME

MAKING TIME

Steve Taylor

Why Time Seems to Pass at Different Speeds and How to Control It

ICON BOOKS

Published in the UK in 2007 by
Icon Books Ltd, The Old Dairy,
Brook Road, Thriplow,
Cambridge SG8 7RG
email: info@iconbooks.co.uk
www.iconbooks.co.uk

Sold in the UK, Europe, South Africa and Asia
by Faber & Faber Ltd, 3 Queen Square,
London WC1N 3AU
or their agents

Distributed in the UK, Europe, South Africa and Asia
by TBS Ltd, TBS Distribution Centre, Colchester Road,
Frating Green, Colchester CO7 7DW

This edition published in Australia in 2006
by Allen & Unwin Pty Ltd,
PO Box 8500, 83 Alexander Street,
Crows Nest, NSW 2065

Distributed in Canada by
Penguin Books Canada,
90 Eglinton Avenue East, Suite 700,
Toronto, Ontario M4P 2YE

ISBN 978-1840468-26-7

Typeset in 11pt Palatino by Marie Doherty

Printed and bound in the UK by
Creative Print and Design Ltd.

To Hugh and Ted

Contents

Introduction

I'm six years old, in the car with my parents and brother, travelling back from our annual two-week holiday in Conway, North Wales. It's dark and the journey seems to take forever. I lie on the back seat, watching the orange streetlights and the houses pass by, and wonder if we're ever going to get home.

'Are we nearly there yet?' I ask my father.

'Don't be silly', he says. 'We only set off half an hour ago.'

My mum plays the 'Yes/No' game and 'Twenty Questions' with us to make the time pass faster. We listen to the radio for a while. Then I fall asleep. When I wake up it seems like I've been in the car for an eternity and I can't believe we're still not home.

'Are we nearly there yet?' I ask again.

'Not far now', says my father.

We play some more games and finally I recognise the streets of our suburb of Manchester. I feel bored and miserable and tell myself that I'm never going to spend as long in a car ever again.

The journey from Conway to Manchester took two hours when I was a child and still takes roughly two hours (although slightly less due to improvements in roads). I made the journey again a few years ago and couldn't believe how short it seemed now, from my adult perspective. Those two hours – which seemed like an eternity when I was six – were nothing. My girlfriend was driving, and we chatted, listened to tapes, watched the Welsh countryside give way to the urban sprawl of north-west England, and we were back in Manchester almost before we knew it. It was a little frightening – what had happened to all the time that two hours contained when I was six years old?

A few years ago I made another journey which gave me an indication of *how* much more quickly time is passing to me now. This was a fifteen-hour plane journey, from Singapore to Manchester, which also seemed to last forever. I'm not a very good flyer and it wasn't a very good flight: we flew into two typhoons over India and it was rocky almost all the way. I hoped I'd be able to 'kill' some of the time by sleeping but it was impossible. Every time I drifted off my anxiety woke me up again. Failing that, I hoped I'd at least be able to make the time pass quickly by distracting myself with in-flight entertainment or books and magazines, but my mind stubbornly refused to move from the moment-to-moment reality of the situation. I was aware of every minute passing, and as a result time seemed to drag horribly. Every time I checked the clock – which was every few minutes or so – less time had gone by than I expected.

My subjective sense of how long that journey took is, I realised recently, very similar to my sense of how long my childhood journey to Conway took. To me they seemed to involve roughly the same amount of boredom and impatience and lasted roughly the same amount of time. This suggests that what was two hours to me as a child was equivalent to fifteen hours to me as an adult – which means, rather frighteningly, that time was passing around *seven times* faster since I had grown up. Where has all the time gone? Why do I seem to have lost so much of it?

About fifteen years ago, not long after I'd left university, I joined a rock group as bass player and toured Germany. While we were over there I fell in love with a girl, and decided not to go back home at the end of the tour. I moved into her flat and ended up staying in Germany for four years. The first few months were a magical time. It was eastern Germany, just a couple of years after the wall had come down, and everything was so different. It was an exhilarating experience just to walk down the street, to see the houses black with coal dust, crumbling to the ground, the strange socialist architecture and the people with the different clothes and hairstyles and language. Everything was incredibly real and exciting – I was in my first serious relationship, making friends with new people all the time, playing in a new band, learning a new language, and starting to teach English. Everyday I was bombarded with newness – new sights, smells, sounds and experiences.

After I'd been there for eight months I went back to Manchester for a holiday and when I arrived, I felt like a Roman soldier returning from years away fighting in a distant country. I couldn't believe it was only eight months since I'd left – it seemed more like eight years. So much time seemed to have passed that I was amazed that everything was still exactly the same, that all the shops were the same with the same people working in them, that all my friends were still doing the same jobs and living in the same houses. Time seemed to have stretched out for me, as if I'd been in a spaceship travelling at close to the speed of light.

This book is my attempt to explain why we experience different perceptions of time such as these. It answers questions which puzzle all of us: Why does time seem to speed up as we get older? Why do new experiences often seem to 'stretch' time, as happened to me in Germany? Why does time seem to 'fly' when we're having fun, or to drag when we are bored or anxious (as in my flight back from Singapore)? And why does it slow down drastically or disappear altogether in accident or emergency situations, in drug experiences, for sportspeople who are in 'the Zone', or for people with mental disorders such as schizophrenia?

Before Einstein, scientists believed that time was something absolute, which flowed at the same rate everywhere in the universe. But Einstein showed that the rate at which it passes depends on two things: speed and gravity. The faster an object moves, the more slowly time passes for it. A clock on a spaceship which was travelling at 87 per cent of the speed of light, for example, would run twice as slowly as a clock on Earth. In a similar way, the more gravity an object has, the more slowly time passes for it. Time runs more slowly on the surface of Jupiter than on the Earth, for example, because of its larger gravitational field.

Einstein dealt with time as it flows in the universe – universal time, if you like. But in this book we're going to investigate *psychological* time, time as we perceive and experience it in our lives. I'm going to sketch out a kind of parallel Theory of Relativity that deals with the inner world rather than the outer, looking at the factors which make time pass at different speeds to us as individuals.

I'll explain that all of our different perceptions of time are based on the same factors working in different ways. By understanding these factors we can learn to control them, and so overcome our sense of time passing too quickly. Our normal state is to feel at the mercy of time – that it's running away from us, never allowing us to do as much as we want to, and taking away our youth, good looks and health. Time is an enemy, eating our lives away and taking us ever closer to death – as the French saying goes: 'The hours are killing you, one by one.'

In addition, the fact that the time is always moving forward means that any happiness we find is only going to be temporary, because the conditions which produce it will soon change. Everything changes – people and relationships change, while success, wealth and good fortune fade away. As the philosopher Schopenhauer wrote: 'In a world like this, where there is no kind of stability, no possibility of anything lasting, but where everything is thrown into a restless whirlpool of change … It is impossible to imagine happiness.'[1]

As a result of this, most of us work hard to hold back the onward flow of time. Some of us exercise and eat healthy foods to try to make sure that we'll live for longer; some of us use moisturiser and beauty products to try to slow down the ageing process, and others have plastic surgery to try to convince themselves (and other people) that they're younger than they are. Many of us also try to preserve our happiness by making our lives as secure and stable as possible – with steady jobs and relationships, routines that never change, insurance policies and pensions – so that 'time the destroyer' can't get at them. (But it always does in the end, of course.)

Time is also our enemy because it seems to work against us in an almost perverse way: it often 'shortens' enjoyable experiences and 'lengthens' miserable or boring ones. Why should a boring afternoon in the office seem to last for an eternity, while an evening out with friends races by?

But my point is that we don't need to feel helpless in this way because we can learn to control the flow of time in our lives. It's actually possible for us to *expand* time, to alter our perceptions so that we experience *more* of it. We don't need to try to cheat the ageing process or extend our lives for as long as possible – it's actually easier, and more beneficial, to expand time from the *inside*, by changing the way we experience the moment-to-moment reality of our lives.

4

We're going to begin by looking at the different perceptions of time we experience in our lives. There are five basic
ones I'll put to you, which I call the 'Five Laws of
Psychological Time'. These obviously aren't 'laws' in the
normal scientific sense of the term – they can't be, since
we're dealing with the realm of the subjective, where there
are no concrete, verifiable and indisputable facts. The laws
I outline should be seen as statements which are generally
true, which fit with most people's subjective experience,
and which have been supported by research. Over the last
four years, I've investigated differing time perceptions and
unusual experiences of time – and I'll refer to my findings
from numerous experiments and interviews.

Then we look at why these different perceptions of time
occur. We'll see that, just as universal time is relative to
speed and gravity, *psychological* time is relative to two factors: information-processing and the ego. By ego, I simply
mean our sense of being an 'I' inside our heads. It's the part
of us which exists as a self-conscious individual and thinks,
plans, makes decisions, daydreams and analyses our experience; it makes us feel separate from the world and from
other human beings. And the ego is, it seems, closely linked
to our sense of time. We'll see how all the unusual perceptions of time we experience can be explained in terms of ego
and information-processing, from the endless days of childhood to the 'timeless moment' of higher states of
consciousness and the slowing down of time which goes
with pain and discomfort.

After this we move on to the second major purpose of
this book, which is to investigate what time actually is, and
to question whether it really exists, at least in the sense we
normally understand it. We're going to look at time from a
cultural and historical perspective, and we'll see that our

normal linear perception of time isn't something absolute, but is particular to the kind of 'psyche' – or consciousness – through which we experience the world. In my book *The Fall*, I suggested that other peoples in the world – and in history – have a different kind of psyche to us, and so experience the world in a different way. And one of these differences is their sense of time. For example, some cultures have a cyclical sense of time rather than linear, while others don't have any concept of time at all. We'll also look at what paranormal, mystical and near-death experiences, together with modern physics, tell us about time. The conclusion we'll consider is that our normal linear sense of time is a kind of illusion, created by our minds. In the light of this, we'll see that not only is it possible to expand time, but also to *transcend* it altogether.

In the final section of the book we'll look at exactly what we can do to control our sense of time passing, to make it speed up or slow down as we like. We'll discuss how to expand the amount of time that we live through, so that we actually experience more time, meaning we live for longer. And we'll also investigate ways to free ourselves from the tyranny of linear time, the 'hurry sickness' which many of us suffer from, the sense that time is running away from us and slowly stealing our lives away, moment by moment. The key to this is *the present*. By giving our attention to the present moment and our present experience – rather than to the 'thought-chatter' inside our heads – the future and the past cease to exist as realities and our sense of linear time is replaced by a powerful awareness of the 'nowness' of things.

Just one point before we start: I don't want to present any of my ideas or theories in this book as ultimate truths. I'm putting them forward in a spirit of debate, as suggestions

go to p. 35

with which you're free to agree or disagree. To me, none of them are fixed; they've evolved as I've been researching and writing the book and they'll no doubt continue to do so. In the same way, it may be that some of the laws or statements about time won't fit with your own experience. Perhaps you feel that time hasn't got faster as you've got older, or that it doesn't go slowly when you're in unfamiliar environments. That's only to be expected since, as I mentioned before, we're not dealing with objective facts or conclusive theories. As you'll no doubt realise, I'm the kind of person who likes playing around with theories and evaluating different answers to the same problem. I've developed these particular theories because, in my opinion, they are the most coherent and cohesive way of explaining the different perceptions of time we experience. But I'm always open to other theories – and if you come up with any of your own, I'm willing to listen.

Just to summarise then, I hope that you'll take away three things from this book. First of all, you'll take away an understanding of time – why it seems to pass at different speeds in different situations and how it's created by our minds. Secondly, you'll gain a control over time in your life, and an ability to expand it so that you can effectively live for longer. And finally, perhaps most importantly, I hope that this book will help to give you what's ultimately the only thing that we have: the present.

1

The First Four Laws of Psychological Time

1. Time speeds up as we get older*

Here's another example of vanishing time, which was given me by a friend. At the age of fifteen he went on a school trip to France. He had the misfortune to go to an all-boys' school, but the special feature of this trip was that girls from the local girls' school were going too. He had a great time, as you can imagine. He drank lots of French beer, smoked French cigarettes, and started a fumbling teenage affair with one of the girls.

A year later he was sitting on a bus and realised that two girls sitting opposite him seemed familiar. After a while it clicked: they were two of the girls who went on the trip with him. He realised that it was almost a year to the day

* Note: if at any point you find the different laws difficult to remember, you can refresh your memory by referring to Appendix 1 on p. 229.

that they went on the trip, and it made him feel nostalgic.
As he told me:

> That year seemed like such an enormously long period
> of time, so long that I'd forgotten what the girls looked
> like properly, even though I spent most of the holiday
> drooling over them. I wanted to go up and speak to them
> but it seemed so long ago that I was afraid they wouldn't
> remember me (although maybe I was just making up
> excuses for my shyness) ... If I had to put it in terms of
> how time is passing for me now, I'd say it was the equiv-
> alent of about four years.

Of course, this first law of psychological time is so familiar
that it doesn't really need to be illustrated with examples.
We've all remarked on it: how Christmas seems to come
round quicker every year; how you're just getting used to
writing the date of the new year on your cheques when you
realise that it's almost over; how your children are about to
finish school when it doesn't seem long since you were
changing their nappies ...

Every time I give a lecture or run a course or workshop,
I present this law and ask if people agree or disagree with
it, noting down the figures and adding them together as a
kind of ongoing survey. And at the moment over 93 per cent
of people I've surveyed feel that time has sped up as
they've got older. In fact the people who disagree with the
law are almost always young people, in their early or mid-
twenties, who presumably haven't yet become aware of a
speeding-up of time. Other, more formal questionnaires by
psychologists have shown that almost everyone – including
college students – feels that time is passing faster now compared to when they were half or a quarter as old as now.[1]
And perhaps most strikingly, a number of experiments have

shown that, when older people are asked to guess how long intervals of time are, or to 'reproduce' these intervals, they guess a shorter amount than younger people.[2]

The two lives

It's sometimes said that human beings live two lives, one before the age of five and another one after, and this idea probably stems from the enormous amount of time which those first five years of our lives contain. It's possible that we experience as much time during those years as we do during the seventy or more years which come after them. As Bill Bryson puts it in his recent memoir of his childhood, *The Life and Times of the Thunderbolt Kid*: 'Because time moves more slowly in Kid World ... [childhood] goes on for decades when measured in adult terms.'[3]

It seems that during the first months of our lives, however, we don't experience any time at all. According to the research of the psychologist Jean Piaget (who conducted a massive number of experiments in order to trace the cognitive development of children), during the first months of our lives we live in state of 'spacelessness', unable to distinguish between different objects or between ourselves and objects. We are fused together with the world, and don't know where we end and where it begins. And we also experience a state of timelessness, since – in the same way that we can't distinguish between objects – we can't distinguish one moment from the next. We don't know when an event begins or when it ends. As the transpersonal psychologist and philosopher Ken Wilber writes, for a newly born child 'there is no real space ... in the sense that there is no gap, distance or separation between the self and the environment. And thus, there is likewise no time, since a succession of objects in space cannot be recognised.'[4]

We only begin to emerge from this timeless realm as our sense of separation begins to develop. According to Piaget, this begins at around seven months. We start to become aware of ourselves as separate entities, apart from the world, and also to perceive the separation between different objects. And as a corollary of this, we begin to be aware of separation between different events. We develop a sense of sequential time, a sense of the past and the future, which is encouraged by the development of language, with its past, present and future tenses. According to Piaget, this process follows four stages. First, we recognise that people arrive and events begin; second, we recognise that people leave and events end; third, we recognise that people or objects cover distances when they move; fourth, we become able to measure the distance between different moving objects or people – and at this point we have developed a sense of sequential time.[5]

After this point of 'falling' into time, we become more and more subject to it. The sense of time speeding up isn't something that we just experience as adults; it probably happens from early childhood onwards. If the sense of sequence is the result of our development of a separate sense of self, we can probably assume that the *more* developed our sense of self becomes, the more developed the sense of sequence will be – meaning that time will move faster. Time may pass for a two-year-old child, but probably only at an incredibly slow speed. But as the child's sense of self becomes more developed, the speed of time increases too. Time probably moves faster to a child of four than it does to a child of three, and faster to a child of seven than it does to a child of six.

However, as my childhood journey to Conway showed, even at this age time passes many times more slowly than it does for adults. This is why, as any parent knows, young children always think that more time has gone by than actually has, and often complain of it dragging. As Bill Bryson puts it, to children time moves 'five times more slowly in a classroom on a hot afternoon, eight times more slowly on any car journey of over five miles'.[6] Primary-school teachers should be mindful of this when their pupils' attention starts to wander – what seems to be a fairly short 40-minute lesson to them (and a fairly short day from 9 am to 3.30 pm) is stretched many times longer to the children. This could have some bearing on childcare practices too. There's been a lot of debate recently about the effects of children being looked after by childminders or nurseries, and one factor which should be considered here is how children *perceive* the time they spend away from their parents. Let's say a parent drops his or her son off at the childminder's at eight in the morning and picks him up at 5.30 pm, and spends two and a half hours with him before putting him to bed. Those nine and a half hours of separation are long enough even from an adult perspective, but for the child they'll feel significantly longer. And this 'stretched out' time could intensify the negative effects of separation, such as a subconscious feeling of rejection, or a weakening of the bond between the child and parents.

Young children's sense of time is undeveloped in other ways too. They can't accurately guess how long events last – in fact they only become able to do this in terms of seconds at the age of six or seven.[7] They don't have a clear sense of the sequence of past events either. When children between the age of two and four talk about what they have done, or retell the story of something that's happened to them, they almost always mix up the order of the events, usually grouping them together in terms of association rather than sequence. Similarly, they generally aren't able to arrange a set of pictures in sequence so that they make a story.[8]

Their awareness of the future is usually very limited too. We tend to forget that the future and the past don't *really* exist. They only exist in our minds. All that really exists is the present, and it's just that, while we're in the present, we have *thoughts* about the future and the past. We remember what's happened to us before the present and we anticipate what's going to happen to us after it. As St Augustine wrote: 'The past is only memory and the future is only anticipation, both being present facts.'[9] And because young children don't have the ability to think abstractly or rationally (according to Piaget, this doesn't begin to develop until the age of 7), they find it difficult to conceive of the past or the future. A couple of years ago my wife decided to take her nephews and nieces to the theatre. That morning she saw her five-year-old nephew Charlie, who was crying for some reason. 'Don't be upset Charlie', she said to him. 'We're going to see *Puss in Boots* this evening.' Charlie just looked at her blankly, as if he didn't understand what 'this evening' meant, and carried on crying. This reminded my wife of one Christmas time when she was a young child. She was having a tantrum and her older brother tried to cheer her up by saying: 'Don't worry Pam, it's Christmas tomorrow – you'll be getting lots of presents.' She remembers that the idea of 'tomorrow' meant absolutely nothing to her. She shouted back: 'But it's not Christmas now!'

According to the German developmental psychologist, Erik Erikson, human beings' sense of time becomes fully developed between the ages of fifteen and sixteen. After living in a state of 'temporal confusion', we now have a clear 'temporal perspective'.[10] This is because, by this stage, our sense of self has become fully developed, including our sense of separation from the world and the ability to think abstractly and rationally. Our 'fall' into time is now complete, but in adolescence and early adulthood our lives are still so full of new experiences that time seems to move fairly slowly. Most people only begin to notice a speeding-up of time in their late twenties or early thirties. By this time we have often 'settled down'. We've settled into our jobs, our marriages and our homes, and our lives have become ordered into routines – the daily routine of working, coming home, having dinner and watching TV; the weekly routine of (for example) going to the gym on Monday night, going to the cinema on Wednesday night, going for a drink with friends on Friday night, etc.; and the yearly routine of birthdays, bank holidays and two weeks' holiday in the summer. After a few years we start to realise that the time it takes us to run through these routines seems to be decreasing, as if we're on a turntable which is picking up speed with every rotation. As the French philosopher Paul Janet noted more than a hundred years ago:

> Whoever counts many lustra in his memory need only question himself to find that the last of these, the past five years, have sped much more quickly than the preceding periods of equal amount. Let any one remember his last eight or ten school years: it is the space of a century. Compare with them the last eight or ten years of life: it is the space of an hour.[11]

Forward telescoping

I'd like you to pause for a moment to answer a few questions. I'd like you tell me the year in which the following events happened.

1. The Lockerbie air disaster
2. The terrorist gas attack in a Tokyo tube station
3. The fall of the Berlin Wall
4. The Bill Clinton/Monica Lewinsky scandal
5. The death of George Harrison of the Beatles
6. The death of Princess Diana

You'll find the answers to these questions at the back of the book on p. 230, and most of you should find that – as long as you're old enough to remember the events – you have, on average, dated these events too recently. This is the phenomenon which psychologists call 'forward telescoping': the tendency to think that past events have happened more recently than they actually have. Marriages, deaths, the birth of children – when we look back at these and other significant events, we're often surprised that they happened so long ago, shocked to find that it's already four years since a friend died when we thought it was only a couple of years, or that a niece or nephew is already ten years old when it seems like only three or four years since they were born. As one 83-year-old man told me: 'I can never guess how long ago things happened. People ask me things like "When did so and so get married?" or "When did so and so die?" and I'm always way out. If I say it was two years it turns out to be five years. If I say six months, it's two years.' The same holds true for national and international events, such as the list above: studies have shown that people usually date these too recently as well.[12] And perhaps this is because time moves more quickly as we get older, with every month and year shorter than the one before. It doesn't seem like four years since a friend died or a baby was born, or since a famous person died, because during those four years time has been speeding up without you realising.

As adults our sense of the past and the future becomes very acute too. Our ability to deliberate and think discursively means that the past and the future become as important to us as the present. Children live in the here and now – they don't worry about whether they're going to get on with others at the birthday party they have to go to next week, or ruminate over what happened to them at nursery last week. Instead they give complete attention to the present moment, to what they're doing, and to the people, objects and other phenomena immediately around them. But as adults we start to live inside our own heads rather than in the world and in the moment – we start to daydream, to deliberate, to worry and to plan. Instead of giving our attention to the here and now we give it the 'there and then', so to speak. We spend so much time considering the future and the past that we become alienated from the present. As Blaise Pascal wrote more than 350 years ago: 'We are so unwise that we wander about in times that do not belong to us, and do not think of the only one that does.'[13]

Meanwhile the turntable keeps picking up speed, compressing every period of time, so that years pass like months and decades like years, until we reach our twilight years and time moves so frighteningly fast that days are over almost while we're still considering what to do with them. One 79-year-old woman I interviewed told me: 'I know I'm old and I've lived for a long time, but I don't feel that old, because the last twenty years or so have just sped by.'

2. Time slows down when we are exposed to new experiences and environments

Here a student describes how slowly time passed for her when she started at university, moving away from her hometown for the first time.

Time was so slow that by the time I went home at the end of the first term it seemed like about two years. So much seemed to have happened in the three months. I felt like I'd been away for so long that I was surprised that everything was the same at home.

It seems that most of us experience a slowing down of time when we leave our normal surroundings and daily routines for a while. As the Buddhist and scientist Jon Kabat-Zinn puts it: 'The inward sense of time slows dramatically when you are off in some unfamiliar place engaged in some adventurous undertaking. Go to a foreign city for a week and do lots of different things, and it seems when you get back that you've been gone for much longer.'[14] New surroundings and new experiences seem to stretch the hours and days longer than normal. When you return home your house seems strangely unfamiliar, and because the week you were away for is much longer than a week of 'home time', you somehow expect to have received massive amounts of post and answering-machine messages, and are always disappointed by the small number of them. The next day you phone your friends and expect them to have lots of news – but, of course, you find that nothing (or at least very little) has happened while you've been away.

On the other hand it may not be a question of expanding time by leaving your familiar environment but of new things happening in your life. I was very aware of this three years ago, after we had our first baby. People I hadn't seen for a while would often ask me: 'So how old's your baby now?' When I told them, almost without fail they'd reply: 'No, he can't be that old! I thought he was –' and they'd mention a far younger age. 'Wow, hasn't the time gone quickly!' they'd say. And I would always reply: 'Not to me it hasn't – to me it seems like about five years since he was born.'

And it was true – when I looked back to how my life was before Hughie was born, it seemed like a different era. Much more time seemed to have passed than actually had. And the reason for this, I believe, is the massive amount of newness which had entered my life since his birth. On the one hand there was the dramatic change to the normal routine of our lives, waking up at all hours, hardly ever going out in the evenings any more, spending all our time looking after this tiny helpless being in a cot at the bottom of our bed. And then there were all the new skills and tasks I had to learn (particularly in the beginning): changing nappies, bathing, changing and feeding him, making up bottles of milk, using the steriliser, and so on. And also the constant newness of watching him develop, as he learned to crawl and walk and then speak, and reacted more and more to his environment. This newness slowed down time; the normal progress of time speeding up as I get older was interrupted by a massive influx of new experience. (Significantly, I haven't noticed this effect so much with our second baby, who's just approaching his first birthday. This is presumably because there has been much less change and newness the second time around.)

It's interesting to speculate that this slowing down of time may be one of the reasons why novelty and unfamiliarity attract us so much, and why we like to go 'away' as much as we can, either for day trips or holidays. The new experience itself gives us a kind of intensity and makes us feel more alive, but it may also give us a sense of freedom from time. In his novel *The Magic Mountain*, Thomas Mann notes how 'periods of change and novelty ... refresh our sense of time ... [During] our first days in a new place, time has a youthful, that is to say, a broadened sweeping, flow.' This, Mann states, is the purpose of all our changes of scenery and trips abroad. But he also notes how this only lasts for a short time – he actually specifies six to eight days – and relates this to familiarity. He writes: 'As one "gets used to the place," a gradual shrinkage of time is felt … until the last week, of some four, perhaps, is uncannily fugitive and fleet.'[15]

A survey at Manchester Airport

In order to put this second law to the test, one afternoon last summer I went with a friend to the arrivals lounge at Manchester Airport. Over the course of two hours, we asked 220 returning travellers – almost all holidaymakers – three questions: 'Where have you been?', 'How long were you there?, and most importantly, 'Do you feel like you've been away for longer than ___ weeks/days?' We often had to clarify the last question by enquiring further: 'Do you feel like it's been a long week/___ weeks?' or 'Do you feel as though time has passed quickly or slowly?' The results did suggest that there was some degree of time-stretching due to unfamiliarity: 61 per cent of people felt that they'd been away for longer than the actual time of their trip, or that it had been a 'long week/two weeks'. On the other hand, 14 per cent said that time had just passed at a normal speed, while 25 per cent said they felt that they'd been away for less than actual time, or that time had passed quickly.

These results do seem to support this second law, even though they perhaps weren't as significant as I expected. But it's important to note that that going on holiday doesn't necessarily mean being exposed to new environments or experiences. A lot of the people we interviewed had returned from 'beach holidays' in resorts – the kind of vacations in which people usually have very little exposure to the actual culture of the countries they visit. In addition, many of the people who said that time had passed quickly or at a normal speed mentioned that they'd been to the place several times before. One lady told me that she'd been going to Malta for her holidays for 25 years; one couple told me that they'd stayed at the same hotel in Turkey every year for nine years. And interestingly, different places did show different results. We spoke to 56 people who'd just returned from Mexico, and 38 (68 per cent) said they felt they'd been away for longer than the actual time. Only 4 (7 per cent) felt the opposite, and 14 (25 per cent) felt that time had passed normally. There were similar results for people who'd just returned from Florida: of 36, 25 (70 per cent) felt they'd been away for longer than actual time. In contrast, we spoke to 44 people who'd just come back from a resort in Turkey, and only 16 (36 per cent) felt they'd been away for longer, with 22 (50 per cent) saying they felt time had gone quickly. A possible reason for this, I would suggest, is that the people who'd been to Mexico and Florida had – on average – more adventurous holidays than the people who'd been to resorts in Turkey, with more travelling and more unfamiliarity. (And indeed, some of the former did tell me that they'd been travelling around.) If you spend most of a holiday sitting on a beach or in bars, relaxing and reading novels and chatting, you're exposed to so little unfamiliarity that there's no reason for time to go any slower than it does when you're sitting in an office or in your living room at home.

Significantly, there also seemed to be a link between how long people were away for and how they perceived time. On average the people who had been away for shorter periods experienced more of a time-stretching effect, supporting Thomas Mann's suggestion that time starts to speed up again after six to eight days in a new place. While most people we spoke to were away for two weeks or more, 62 of them were away for periods of between two and a half and eleven days. And a significantly higher number of these (71 per cent) felt that time had passed slowly. Presumably, as Mann argues, this is because a new place starts to become familiar after a week or so. In fact, several people said as much to us, remarking that: 'The first week seemed to go slowly, then it went fast' or 'I felt like I'd been there for a long time and then the last few days sped by.' Other people made interesting comments like: 'When you're there it's slow, but when you're back it's too quick' and 'When I go on working trips abroad, I feel like I'm away for a long time, but when I'm relaxing on holiday it doesn't seem as long'.*

The two twins

One of the interesting things about this second law of psychological time is that it shows how different two people's perceptions of the same period of time can be. Here we can use the 'two twins' metaphor, which is often used to illustrate universal relativity. Let's say there are two 30-year-old twins, one of whom decides to give up his job and go travelling around the world for a year, while the other carries on living his normal life, working in an office and looking after his children. For the twin who goes away, that year will, of course, contain a massive amount of time; while for the one

---

*The ideal experiment to test this law would be one which included a kind of 'unfamiliarity index', a way of grading people's trips in terms of the new experiences they are exposed to. However, it's difficult to see how this could be done.

who stays at home, it will pass quickly. So when they see each other again the first twin might say 'It's so long since I've seen you, I can't believe it's really you', while the other twin might say 'It only seems like yesterday that you left'.

In other words, this law shows that it's possible for two people to live through different amounts of time in the same time period. The twin who goes away experiences more time in that one year than his brother does, he lives through more time, so in a sense, he has now actually lived for longer than his brother. As we'll see later, this has important implications when we think about how 'long' our lives actually are, and how it's possible for us to control time.

I'm aware that all of these judgements about time are *retrospective* ones, of course. We're talking about becoming aware of how slowly time has passed at the *end* of the day, the week or the holiday. But what about our sense of time passing *now*, in the present, you might ask? Is it going slow or fast for us at this particular moment? But if you think about it, we can *only* be aware of the passage of time in retrospect. It's impossible for us to say that time is going slowly or quickly for us at a particular moment, because we don't have a sense of time moving, we only have a sense of a static present. We flow with time, we're *in* it, and so can't measure it. We can only do this when we stop and look back at what has just passed. Time really *does* move slowly for you during your holiday or your weekend away, just as it really does move quicker for older people – even if you only become aware of this retrospectively.

3. Time passes quickly in states of absorption

Here a young man I interviewed describes how his perception of time changes when he plays on his Gameboy.

Time just vanishes. I get so absorbed that I completely forget about time. And when I come out of it I'm always shocked at how much has gone by. It's frightening – hours seem to pass by in minutes. Once I played on it for twelve hours and if someone asked me to guess I would've said that it was only three or four.

And here, similarly, an acquaintance of mine describes an experience of time passing quickly as a result of a similar degree of absorption. She had booked a coach from Northern Malaysia to Singapore. It was a twelve-hour journey, and she was dreading it. Apart from the sheer boredom and discomfort, she'd been ill with a migraine and was prone to travel sickness. But amazingly the journey turned out to be fairly pleasant.

Luckily there was a video player on it [the coach]. After a while they put *Titanic* on; I'd never seen it before and got completely absorbed in it. I forgot about time and it's a very long film, about four hours, so the next thing I knew we weren't far off halfway through the journey and it seemed like we'd just got started. Suddenly the journey didn't seem so long after all. They put another film on, and I got absorbed in that too; and by the time it was over we didn't have that much longer to go. The films put me into a sort of 'timehole' which made all the time pass by without me being there.

However, absorption doesn't just mean entertainments like television or computer games, it also – perhaps mostly – means *being busy*. It means being engaged in an activity which occupies your attention and stops you being aware of your own thoughts and your surroundings, whether it's making a meal, writing an email or running a newspaper shop.

A couple of years ago I was standing at the checkout of a very quiet supermarket, as a glum-looking girl ran my shopping through the barcode machine. 'It's quiet today, isn't it?' I said to her. 'Oh yes', she replied. 'It's terrible. I hate it when it's quiet – time drags so much.' At the time I was surprised by her reaction, assuming that if you have to do a monotonous job you'd prefer not to be busy. But I didn't take into account the time-slowing effect of inactivity. This was brought home to me just a few days ago in another supermarket, a few minutes before it was due to close. The woman at the checkout seemed very cheerful and so I asked her: 'Have you had a good day?' 'Oh yes', she said. 'It's been so busy. The time's gone so fast. When it's really busy you just look up and four hours have gone by.'

Most of us spend a large part of our lives in what you could call a medium state of absorption, with our attention immersed in activities or entertainments. We spend eight hours a day with our attention immersed in our jobs (to greater or lesser degrees, depending on how demanding and interesting they are); we have tasks to do in our free time which absorb our attention, such as housework and DIY; and we also spend a lot of our free time with our attention immersed in TV programmes, or in other kinds of entertainment such as magazines or games. But because this medium level of absorption is so normal, it's usually only when we experience *intense* degrees of absorption that we notice the effects of this law. Computer games are well known for the time-devouring effect they have. They demand more personal involvement and concentration than any other kind of home entertainment, and so produce a more intense degree of absorption. But we also get into states of intense absorption when we watch particularly enthralling TV programmes, films or sports matches or read especially good books. In these situations we forget our-selves, forget our surroundings and forget time. We seem to go outside time; or to use the last interviewee's analogy, slip into a timehole, which enables us to not 'be there' while time passes. It passes without us and when we rejoin its flow we are always surprised at how far it's moved on.

These effects have been verified by a large number of experiments by different researchers. These have estab-lished that, in the language of academic psychology: 'When an interesting stimulus or a stimulus that requires more attentional resources is presented during the interval to be estimated, fewer time units are processed and the individ-ual tends to underestimate the temporal intervals.'[16] In other words, the more intensely you pay attention to some-thing in a particular period of time, the shorter you think that period of time is. In one study by Hawkins and Telford, participants listened to different prose passages on tape and experienced the more interesting ones as shorter than the boring ones. And a study by Sawyer et al. in 1994 found that when participants were asked to do more complex tasks (which required more attention) they also experienced less time.[17]

In one of my own experiments, I split my lecture audi-ence of 30 people into two groups. I gave the people in the first group a sheet with the numbers 1–10 in Chinese (writ-ten phonetically), and asked them to try to learn them. Meanwhile I asked the people in the other group to do noth-ing – to literally just stare out of the window without talking, and let their minds wander. The exercise lasted for 4 minutes 30 seconds, and at the end I asked both groups to estimate how much time had passed. The people who did nothing estimated an average of over 50 seconds longer. Their average estimate was longer than the actual time, whereas the numbers group was slightly shorter. I have repeated this experiment several times and always had a similar result. The people who learn the numbers are in a state of absorption – to a greater or lesser degree – and so time goes faster for them.

In another experiment I split a group of students into two groups of seven. Telling them simply that we were going to read some materials about time to discuss, I gave one group an entertaining and readable article and the other a highly academic and verbose academic paper. At the end of the period I asked them first to write down how much time they felt had passed, and also to rate the readability of the articles on a scale of 1 to 10. The average estimate of the group who had the academic paper was longer, by more than a minute. The rating of the readability of the academic paper was lower too – an average of 4 compared to 7 for the article. There was a clear relationship between how interesting the students found the articles, and how much time they experienced. Less interest equals more time.

Active and passive absorption

As well as different degrees, there are two different *kinds* of absorption, which you could call passive and active absorption. Passive absorption is absorption without concentration, when we don't actively *focus* our attention – the state we sometimes get into when we watch television, for example. Active absorption is absorption *with* concentration, when we do focus our attention, such as when you do difficult and challenging tasks at work, play a musical instrument, repair a car, or write a story.

Psychologically, there's a great deal of difference between these two kinds of absorption. Active absorption (or concentration) usually brings a sense of well-being and makes you feel more alive and energised. Passive absorption (or distraction) usually has the opposite effect – it leaves you feeling a little unfocused and devitalised, as if your energy has been drained away. In terms of time though, there doesn't seem to be any difference between them – time seems to pass just as quickly when you're actively engrossed in a game of chess or painting a picture as it does when you passively watch television. The psychologist Mihaly Csikszentmihalyi has made a study of states of 'active absorption', which he calls 'flow', and one of his findings is that in them: 'Often hours seem to pass by in minutes; in general, most people report that time seems to pass much faster'.[18] The American anthropologist Edward T. Hall (in *The Dance of Life*) also notes how concentration makes time pass faster, and gives the example of a 'talented filmmaker from Switzerland [who] experiences the time between meals when she is working as mere minutes. Suddenly she will feel hungry and realise hours have gone by.'[19] Or as a chess player told the researcher Rhea A. White, when he is intensely concentrating on a game: 'Time passes a hundred times faster. In this sense it resembles the dream state.'[20]

Killing time

We all put this law of psychological time to our advantage, when we have long periods of time in front of us which we don't know what to do with, or situations which we think we're going to find painfully boring. If you have to go on a big journey, endure a long wait at a hospital or the dentist's, or even if you're at home for an evening with nothing in particular to do, your first instinct is to find something to take you out of the moment-to-moment reality of the situation and make the time pass quicker. At the dentist's you flick through magazines or do a crossword, on the tube to work in the morning you listen to your iPod or read your book, if you have to stay at home in the evening you rent a video – and this often isn't so much because you're interested in these things in themselves, but because you want to slip outside time, to escape the dreary minutes or hours ahead of you, or at least make them pass faster. In fact we often call the activities which help us to do this 'pastimes' – i.e. activities which help to make time pass faster.

This is what we mean when we talk about 'killing time' – we use distractions to try to erase unwanted periods of time from our days. In a way this doesn't make any sense. We want to live for as long as possible; we try to keep fit and eat good food to *extend* our lives, but when we kill time we're actually *shortening* our lives, by making the time in our life pass faster. (We're going to look at this paradox in more detail in the next chapter.) On the positive side though, you could say that this is one way in which to take control of our perception of time, and make it pass the way we'd like it to, rather than being at its mercy.

This law of psychological time is also reflected in the common saying that 'time flies when you're having fun'. To a large extent, fun *is* absorption. The fun or enjoyment we have when we play or watch sports, listen to music, or chat to friends – in these moments our attention is always absorbed by an activity or spectacle, which means that time passes quickly. When we say 'time flies when you're having fun' we really mean 'time flies when you're absorbed'.

Incidentally, there might be a little confusion here between this law and the previous one ('Time slows down when we are exposed to new experiences and environments'). After all, don't we experience a state of absorption when we go to a foreign country? Aren't we absorbed in all the unfamiliarity there in the same way that we're absorbed in a computer game or writing a story? If so, why is that time goes slowly in the former case, but quickly in the latter?

It's simply a question of perceptual information. When you watch TV or play a computer game your attention is so intensively focused that you screen out the rest of the phenomenal world. Your absorption actually prevents you absorbing any perceptual information (except from the TV show or the computer game itself). But when you're surrounded by unfamiliarity your attention isn't just focused on one particular thing; it's open to everything, a wide-ranging spotlight of attention rather than just a narrow beam, and this allows you to process a large amount of perceptual information. In fact, in my view, absorption isn't really the right word for the state of mind we experience in these moments, since the term implies concentrating your attention on one particular object. It's probably better to use the term attentiveness, or the Buddhist term mindfulness.[21]

4. Time passes slowly in states of non-absorption

It's one of the cruel ironies of time that it passes quickly when we'd like it to pass slowly (as in the third law of psychological time) and that it passes slowly when we'd like it pass quickly. Almost as if some malevolent god is playing tricks on us, not only does it fly when we're having fun, it seems to drag when we're bored and unhappy.

The fourth law of psychological time is the exact opposite of the third. Just as time goes quickly when we're in states of absorption, it goes slowly when we're in states of *non*-absorption – in other words, when we don't focus our attention on anything around us, and our minds aren't occupied. One of my own most painful experiences of this was my fifteen-hour flight from Manchester to Singapore, which I mentioned in the Introduction. Because I was nervous and uncomfortable and couldn't become absorbed in my books or the in-flight entertainment system, I had no distraction from the moment-to-moment reality of my predicament, and so time passed incredibly slowly. And here, as another example, a man I interviewed describes how slowly time dragged when he was doing a boring job and couldn't wait for the end of each day.

> The mornings weren't so bad but the longer the day went on the slower time seemed to pass. By the time it got to four o'clock I was looking at the clock every couple of minutes and every time I looked I just couldn't believe how little time had gone by. I never got used to it, no matter how many times it happened – I always thought the clock must have stopped or was going slow.

This law of psychological time is the reason why time seems to go so slowly when we're reading a dull report, listening to a dreary lecture, or having a boring conversation with a person we don't really like. It's also the reason why time often goes so slowly when we're *waiting*. As we've just seen, waiting isn't a problem for us if we can make our attention absorbed in something. However, it's sometimes difficult for us to do this, either because we feel too anxious to be able to distract ourselves (e.g. before a job interview), or because we have to keep our attention focused on the thing we're waiting for, such as the bus coming around the corner.

The third and fourth laws of psychological time are the basis of the old folk-saying 'busy time goes by quickly and empty time goes by slowly'. Busy time means that your attention is focused on an activity or a distraction; empty time means that your attention is unfocused, and your mind is unoccupied.

Just as the more absorbed we are the faster time passes, the more bored or unabsorbed we are the slower it seems to pass. In fact, absorption and non-absorption are really just different ends of the same spectrum. If you think of boredom as the negative half of a spectrum and absorption as the positive half, time moves slowest at the most intense states of boredom, and picks up speed as you cross over into the positive side of the spectrum and increase the degree of absorption.

And here we can see another example of how the same period might last different lengths of time to two different people. Imagine two students in the same hour-long lecture at college, one of whom finds it interesting and the other finds it boring. The first student keeps her attention fixed on the lecturer and busily makes notes, while the other stares into space and thinks about what she's going to do that night. The lecture lasts the same amount of clock-time for both of them, of course, but the second student actually experiences more time in the hour than the first.

Or imagine two employees in the same office one Friday afternoon. The manager is sitting at his desk, finishing an important report which he's been working on for a week, which he hopes will improve his chances of promotion and make a difference to the company's working methods. Then there's the bored and frustrated office clerk, standing at the photocopier with a giant heap of papers, glancing at the clock every few minutes. To the manager the afternoon whizzes by, of course, whereas to the clerk the same afternoon passes agonisingly slowly.

You can extend this to our whole lives too. Returning to the 'two twins' idea, let's imagine that one of them spends his working life doing a job which is mind-numbingly dreary and doesn't require any kind of mental effort, while the other spends his doing one which is challenging and demanding. Assuming they do similar things in their leisure time (and that they live in the same 'ordinary' state of consciousness), there will actually be more time in the first twin's life than the second's. Assuming they live for roughly the same amount of calendar time, the first twin will actually experience more time in his life than the second. In a sense, then, he lives for longer. (The downside of this though, is that the first twin's life will probably be much less happy and more unfulfilling than the second's, and his life will also probably be shorter in terms of calendar time, since boredom often leads to psychological turmoil and a reduced resistance to disease.)*

These first four laws certainly aren't exclusive. There may be situations when two or more of them operate at the same time, so that they work against each other. For example, although it's generally true that we experience a speeding-up of time as we get older, there may some situations (such as the birth of a baby, in my own life) when the second law (new experiences) comes into play and stretches time. There may also be a situation in which you're exposed to newness (making time pass slowly) while also absorbing

---

* Incidentally, at one my lectures a gentleman asked me why, since I had included the fourth law which is the opposite of the third, I hadn't also included an opposite of the second law. This would have been something like: 'Time speeds up when you are in a familiar environment and have familiar experiences.' A good question, but I didn't feel this should qualify as a law in itself, simply because we spend the vast majority of our lives in familiar environments having familiar experiences. This means the passing of time in this situation is our *normal* experience of time, and is not quicker than usual.

information (making it pass quickly): e.g. if you're on holiday in an unfamiliar environment but spend a lot of your time reading novels and watching TV.

Dealing with any kind of problem or issue usually involves three different stages: first, you identify the problem; second, you find its cause, and third, you try to find a solution to it, based on controlling the cause. We've spent this chapter doing the first of these, and in the next chapter we're going to try to deal with the second of them, by trying to establish the *causes* of these four laws of psychological time. (As I mentioned in the Introduction, there are five laws of psychological time, but since the fifth stands apart from the others in some respects, and is more complex, I'm going to deal with it separately in Chapter 4.) In the next three chapters, I'm going to present a kind of 'Unified Theory of Time', which relates the laws of psychological time to each other. It's possible to do this because, as we'll see in the next chapter, all the different perceptions of time we experience are caused by the same basic factors, working in different ways.

2

How Information Stretches Time: The First Two Laws Explained

It isn't just children's perception of time that's different to ours, but also their perception of the world around them. The world is much more *real* to them, a much brighter, more fascinating and beautiful place.

When I go walking with my three-year-old son Hughie through the fields and paths close to our home, I'm always amazed at how long it takes us to get anywhere. What should be a 10-minute walk by the golf course to the nearest post office can last anything up to 40 minutes. This isn't just because his small legs mean that he's a slow walker – it's mainly because he stops every few seconds to examine something. Trees, bushes, stones, leaves, wire fences, puddles, even discarded crisp packets and coke cans – everything is a source of wonder to him. His world is filled with fascinatingly different textures and colours and shapes and patterns and smells and sounds. Hughie can spend 10 minutes examining a leaf, staring at it, stroking it, brushing it against his face. One of the reasons why it's always so difficult to get him out of the bath is because he loves to just sit there and pour water down from a cup, transfixed by the bubbles and splashes and ripples.

In fact this is one of the great things about having children: it rekindles something of that fresh vision of the world in you. Normally I walk to places like an arrow heading to its target – focused on my destination, paying little attention to my surroundings, my mind on other things. But walking with Hughie has reminded me to stop and look – that almost everything is fascinating if you just take the trouble to pay attention to it. I've realised the joys of just ambling along, staring at the sky, looking at the plants and bushes and trees around me, taking in the reality of the moment rather than thinking about the future or past. (Although I can't pay *too much* attention to all of this, of course – at the same time I have to keep half an eye on Hughie, in case he steps in some dog dirt or decides to crawl through a hole in the fence.)

In general, children live *in the world* much more than adults do. We always live in the world in a physical sense, of course, but generally in a mental sense we don't, because we usually pay very little attention to our surroundings. When was the last time you really *looked* at the streets and buildings you go past on the way to work in the morning? When was the last time you looked at the strange cloud formations above you while you were walking to the shops? When was the last time you really looked at the trees you pass every day?

Rather than being focused on the phenomenal world, our attention is usually either focused inside us, on the thoughts and daydreams which run through our minds, or else fixed to external stimuli such as tasks, activities or distractions. On the train we're much more likely to be thinking about what we have to do at work that day, or to be reading the newspaper or writing text messages, rather than staring out of the window at the towns and the countryside passing by. On a ferry journey most of us are likely to be drinking in the bar or watching TV in the lounge rather than up on deck watching the swell of the sea and the land disappearing in the distance. Of course, we give *some* of our attention to the world at the same time as giving it to distractions or to our own thoughts, but usually only a very small part – just enough to keep us walking in the same direction or to stop us bumping into things.

But children are the opposite. For a lot of the time they give their *full* attention to the world. They love to sit by the window on train journeys so they can stare at the scenes passing them by, they freeze with awe when they see animals in the zoo, and they're forever pointing and shouting 'Look!' at all kinds of phenomena which seem mundane to adults. In fact it's partly because they live 'in the world' to such an extent that children find it so difficult to concentrate. They'll start to do something and then break off when another strange and fascinating sight or sound comes along and takes away their attention. As the psychologist Ernest Becker writes, in childhood we experience a 'vision of the primary miraculousness of creation', and our perceptions of the world are 'suffused ... in emotion and wonder'.[1] The developmental psychologist Alison Gopnik notes that babies and young children have an 'infinite capacity for wonder', and suggests that we only approach their intense vision at our highest moments – for example, when a scientist is inspired by the wonder of the physical world, or a poet is awestruck by beauty.[2]

This is one of the reasons why we often recall childhood as a time of bliss – because the world was a much more exciting and beautiful place to us then, and all our experiences were so intense. Gradually, as we get older, we lose this intensity of perception, and the world becomes a dreary and familiar place – so much so that we stop paying attention to it. After all, why *should* you pay attention to the buildings or streets you pass on the way to work? You've seen these things thousands of times before, and they're *not* beautiful or fascinating, they're just ... ordinary. As Becker describes it, we 'repress' this intensity of vision. 'By the time we leave childhood', he writes, 'we have closed it off, changed it, and no longer perceive the world as it is to raw experience'.[3] Or as Wordsworth puts it in his famous poem 'Intimations of Immortality', the childhood vision which enabled us to see all things 'apparelled in celestial light', begins to 'fade into the light of common day'.[4]

The summer before last we stayed in a cottage in Cornwall for a week, and while we unloaded the car Hughie sat down on the gravel drive and started playing with the stones. He must have sat there for over half an hour, examining each stone individually as if it was a precious jewel. As I walked past him with our bags I caught myself thinking: 'Isn't it amazing that babies are so fascinated by such mundane things?' Straight away I realised the stupidity of the thought, and when the car was empty I sat down with him on the gravel. I picked up some of the stones, looked at each one in turn and realised that he was right to spend so much time with them. They *were* fascinating. Rather than just being a plain grey colour, as I'd assumed, they were made up of lots of different materials, some of them a shiny silver colour and others a transparent quartz. Some of them had a cold and smooth texture, others were rough and jagged. They were beautiful, glistening in the sunlight, all with their own intricate pattern and shape. I felt as though I had caught something of the 'raw experience' of the world which Hughie always perceives, and managed to reawaken some of the 'celestial light' of childhood.

Why time speeds up – the 'Proportional Theory'

The fact that children are so 'awake' to the world is, I believe, part of the reason why time is so expanded to them. But before we go into this in detail, I'd like to look at some other theories of why time seems to speed up as we get older.

In Chapter 1, I suggested that time speeds up for children because of the development of their ego, but this can't explain why time keeps getting faster for us as adults, because by the age of fifteen or sixteen the development of our egos is complete. There's clearly another factor (or other factors) which takes over from this as we enter adulthood.

One popular theory of why the speed of time increases is that, as you get older, each time period constitutes a smaller fraction of your life as a whole. This theory seems to have been first put forward in 1877 by Paul Janet. As William James, the 19th-century psychologist, described it: 'The apparent length of an interval at a given epoch of a man's life is proportional to the total length of the life itself. A child of 10 feels a year as $1/10$ of his whole life – a man of 50 as $1/50$, the whole life meanwhile apparently preserving a constant length.'[5] At the age of one month, a week is a quarter of your whole life, so it's inevitable that it seems to last forever. At the age of fourteen, one year constitutes around 7 per cent of your life, so that seems to be a large amount of time too. But at the age of 30 a week is only a tiny percentage of your life, and at 50 a year is less than 2 per cent of your life, and so your subjective sense is that these are insignificant periods of time which pass very quickly.

There is some sense to this theory – it does offer an explanation for why the speed of time seems to increase so gradually and evenly, with almost mathematical consistency. One problem with it, however, is that it tries to explain present time purely in terms of past time. The assumption behind it is that we continually experience our lives as a whole, and perceive each day, week, month or year becoming more insignificant in relation to the whole. But although our thoughts often focus on the past, we don't actually *live our lives* in awareness of the whole of our past lives. We live in terms of smaller periods of time, from day to day and week to week, focusing mostly on the recent past and the near future, and largely independent of what has gone before.

Biological theories

You can also look at this speeding up of time from a biological perspective. One theory is that our sense of time changes with age because our metabolism slows down as we get older.[6] This makes some sense too, since our metabolism moves fastest when we're children, and seems to gradually and evenly 'wind down' as we grow older – which also fits with the gradually and evenly increasing speed of time which we experience. Because children's hearts beat faster than ours, because they breathe more quickly and their blood flows more rapidly, etc., their body clocks 'cover' more time within the space of 24 hours than ours do. Children live through more time simply because they're moving through time faster. Think of a clock which is set to run 25 per cent faster than normal time. After twelve hours of normal time it has covered fifteen hours, after 24 hours of normal time it has covered 30, which means that, from that clock's point of view, a day has contained more time than usual. On the other hand, old people are like clocks which run slower than normal, so that they lag behind, and cover less time than a normal clock in 24 hours.

In the 1930s the psychologist Hudson Hoagland conducted a series of experiments which showed that body temperature causes different perceptions of time. Once, when his wife was ill with the flu and he was looking after her, he noticed that she complained that he'd been away for a long time even if he was only away for a few moments. With admirable scientific detachment, Hoagland tested her perception of time at different temperatures, and found that the higher her temperature, the more time seemed to slow down for her, and the longer she experienced each time period. Hoagland followed this up with several semi-sadistic experiments with students, which involved them enduring temperatures of up to 65°C, and wearing heated helmets. Results showed that raising a person's body temperature can slow down their sense of time passing by up to 20 per cent. And the important point here, as the scientist and author E.W.J. Phipps notes, is that children have a higher body temperature than adults, which may mean that time is 'expanded' to them. And in a similar way, our body temperature becomes gradually lower as we grow older, which could explain a gradual 'constriction' of time.[7]

However, as we'll see in a moment, the real problem with these theories is that, although they work quite elegantly as far as this first law of psychological time is concerned, they don't throw any light at all on the other four laws. It's much more likely that all the time-distortions we experience have the same fundamental cause.

The 'Information Theory'

Other theories suggest that the first law of psychological time is connected to perception and experience. The British psychologist John Wearden, for example, believes that our sense of time speeding up is a kind of retrospective illusion created by our memory. His research suggests that our memories tend to 'shrink' periods of time in which very little happens, whereas it 'magnifies' hours or days in which a lot happens. As we grow older our lives usually become more monotonous and less exciting. We do fewer things, and fewer things happen to us. Each passing year contains less experience, and so our memories shrink the years, and we feel that time is moving faster. Or as Wearden puts it: 'You find yourself thinking "It's only event 17,500 of the year and yet already it's Christmas. Normally I should be on event 25,000 by now."'[8]

Towards the end of the 19th century, William James put forward a similar view in his classic work, *The Principles of Psychology*.

> In youth, we have an absolutely new experience, subjective or objective, every hour of the day. Apprehension is vivid, retentiveness strong, and our recollections of that time, like those of a time spent in rapid and interesting travel, are of something intricate, multitudinous, and long drawn-out. But as each passing year converts some of this experience into automatic routine which we hardly note at all, the days smooth themselves out in recollection to contentless units, and the years grow hollow and collapse.[9]

James makes the same connection between time and experience as Wearden, but for him it's not so much a question of the *number* of experiences, but the *newness* of our experience and the intensity of our perception. Time goes slowly for children because all their experiences are new and because they perceive the world with a heightened intensity. But as we grow older our experience becomes routine, and our perceptions become less vivid.

For both Wearden and James, the sense that time speeds up isn't something which actually happens to us in the present; it's something that we only become aware of in the future, when we look back at our lives. In other words, newness doesn't make time pass slower for children *while* they're children; the important thing is the larger amount of memories the newness leaves us with, which give us the *impression* that time was passing slower when we look back in later life. As James suggests, 'a week of travel and sightseeing may subtend an angle more like three weeks in the memory … The length in retrospect depends obviously on the multitudinousness of the memories which the time affords.'[10]

However, it's also possible that this speeding up of time – and its slowing down when we go travelling abroad – isn't just a memory, but a direct perception of the present. It's true that we can only judge how time has passed in retrospect, but that doesn't necessarily mean that the sense that time speeds up is a purely a retrospective phenomenon. It's possible that that's exactly how we experience time as we live our lives – a child actually *does* live through more time than an adult, and a person on holiday actually *does* live through more time than a person in their home environment. When we look from the future, the reason why we view eventful periods as lasting longer is not because of the memories they have created, but because that was exactly how we experienced them.

Psychologists' time-estimation experiments argue against the 'memory' interpretation as well. As we saw in Chapter 1, when people are asked to estimate how much time has passed, or to indicate how long different periods of time are, older people consistently give lower estimates than younger people. They are apparently experiencing less time in those periods than younger people. And if a 5-minute period passes more quickly to an old man than a young person, this suggests that it's not his memory which is making time speed up, but that time is actually moving quickly for him *now*, as a real phenomenon.

In my view, the speeding-up of time we experience is related to our perception of the world around us and of our experiences, and how this perception changes as we grow older.

The speed of time seems to be largely determined by how much perceptual information our minds absorb and process – the more information there is, the slower time goes. This connection was verified by the psychologist Robert Ornstein in the 1960s. In a series of experiments, Ornstein played tapes to volunteers with various kinds of sound information on them, such as simple clicking sounds and household noises. At the end he asked them to estimate how long they had listened to the tape for, and found that when there was more information on the tape (e.g. when there were double the number of clicking noises), the volunteers estimated the time period to be longer. He found that this applied to the *complexity* of the information too. When they were asked to examine different drawings and paintings, the participants with the most complex images estimated the time period to be the longest.[11]

Interestingly, Ornstein also found that time seems to contract as we get used to stimuli. In one experiment, for example, he found that when volunteers were shown a series of images in an orderly sequence, they estimated the time to be shorter than others who were shown the same images in a *random* sequence. He suggests this is partly because the students who were shown the 'easily-coded' images stopped responding to them after a short while, because they quickly became familiar, while the others couldn't 'switch off' because they didn't get used to the images. As Ornstein concludes in his book *On the Experience of Time*: 'When an attempt is made to increase the amount of information processing in a given interval, the experience of that interval is lengthened.'[12]

I tested this myself with a simple experiment with music. During a course, I asked the participants to listen to two pieces of music. One was a mad, frenetic piano concerto by Rachmaninov, with notes cascading at a rate of about ten per second. The other was a piece of ambient music by Brian Eno, which floated gently and sedately across the room. We listened to the pieces for different periods of time and I asked the participants to estimate how much time had passed. If time perception is related to information, they should have experienced more time for the Rachmaninov piece. It contains a lot more information than the Brian Eno piece – many times more notes, tones and different instruments. All of this extra information should have stretched time.

And this is what the results showed. We listened to the Rachmaninov piece for 2 minutes 20 seconds, and the average estimate was 3 minutes 25 seconds. We listened to the Brian Eno piece for 2 minutes, while the average estimate was 2 minutes 32 seconds – still an overestimate, but a lower one.

And if more information slows down time, perhaps part of the reason why time goes so slowly for children is because of the massive amount of 'perceptual information' that they take in from the world around them. Their heightened perception means that they're constantly taking in all kinds of details which pass us adults by – tiny cracks in windows, insects crawling across the floor, patterns of sunlight on the carpet, and so on. Even the larger-scale things that we do also observe seem to be *more* real to them, to be brighter, with more presence and 'is-ness'. All of this extra information stretches out time for children. And, in the same terms, perhaps it speeds up for us because, as we get older, we lose this freshness of vision. We begin to 'switch off' to the wonder and is-ness of the world, and gradually stop paying conscious attention to our surroundings and experience. As a result we take in less information, which means that time is less stretched and passes more quickly. The longer we're alive, the more familiar the world becomes, so that the amount of perceptual information we absorb decreases with every year, and time seems to pass faster.

Switching off to reality

It's important to pin down *exactly* why we might process less information as we get older.

There seems to be a kind of mechanism in our minds which 'desensitises' us to things once we've been exposed to them for a while, making our experience less intense and real. Think about what happens when you go into a smoky room, for example. At first you find it unbearable, but after a few minutes you start to get used to it, and when a new person comes into the room after half an hour or so and says 'It's so smoky! How can you stand it?', you're surprised at their extreme reaction. Or think about what happens when you move into a new house in a different area. At first your new surroundings seem intensely real, you're aware of every new sight and sound and feel excited by the strangeness – but after a few weeks the newness begins to fade, and the environment becomes familiar and even dreary.

The same thing happens if you spend time in a different country. As we've already noted, time often seems to pass very slowly during the first week of a holiday – but then the environment starts to become familiar, you start to 'switch off' to its newness, and as a result time seems to speed up. I had the same experience when I went to live in Germany. As I described in the introduction, at first there was so much newness and freshness in my surroundings and in my life that for months I was in a state of exhilaration. But eventually I became desensitised to the environment, and started to feel exactly the same way about my new hometown as I had about the dreary English town I'd left. All the sheen had disappeared from my experience there, it didn't stimulate me anymore, and instead I began to feel the exhilaration of strangeness and newness when I went back home to England for holidays.

Something similar happens when we 'get used to' things in our lives and start taking them for granted. Think of what might happen when a man is released from prison after five years, for example. Psychologists have found that released prisoners go through a number of different stages. First they feel a sense of euphoria and freedom. You can imagine how fantastic it feels for them just to walk the streets and look around at the trees and the sky after staring at four walls almost all of the time, and to visit friends they haven't seen and places they haven't been to for years. But after a certain amount of time – usually anything from a few hours to two weeks – disillusionment sets in. The intensity of their experience starts to fade away, and their sense of freedom is replaced by depression and alienation, as they miss the prison routine and their friends there and realise how difficult life outside is going to be. After a certain amount of time the person begins to adjust and starts to feel more confident, although this may be counteracted by the difficulty of being accepted by mainstream society again – which might lead to a return to crime, and a return to prison.[13] Or perhaps you've had a similar experience after recovering from serious health problems – for a short time you're fully aware of the value of your health, you feel incredibly grateful to be able-bodied and free of illness, but then you start to switch off to the reality of the situation and your health no longer seems such a blessing.

This is such a common feature of our experience that it seems completely inevitable. But it's important to realise that the change in these situations – when something unfamiliar and intense becomes familiar and 'normal' – happens in our minds. The process of getting used to things is the result of a psychological change, when this desensitising mechanism starts to function.

The whole point of this mechanism seems to be to conserve energy. If all of your surroundings and your experience are intensely real to you, you spend a lot of time attending to them, taking in their reality, which uses up a great deal of energy. A few thousand years ago, when life was much more difficult, this would have been a definite disadvantage for our ancestors. It would've been more beneficial for them to devote their mental energy to the practical tasks of survival. They needed to think and deliberate – and concentrate – more in order to survive. So perhaps this desensitising mechanism developed as way of enabling them to do this. It edited out the is-ness of their surroundings so that they wouldn't 'waste' mental energy in perceiving them. It made their experience less intense so that they wouldn't focus their attention on it, and so enabled energy to be redirected to the ego which thinks and deliberates.

It's important to note that this applies specifically to *our* ancestors. As I point out in *The Fall*, most of the modern peoples of Europe and Asia (and their descendants in colonised countries) are descended from groups who originally lived in central Asia and the Middle East. For example, the majority of Europeans are descended from the Indo-Europeans, who archaeologists believe were originally from the steppes of southern Russia, while many of the inhabitants of the Middle East are descended from Semitic peoples originally from what is now Saudi Arabia.

There's a great deal of evidence to suggest that some prehistoric peoples – and until recently, many contemporary indigenous peoples – didn't have a particularly hard time keeping themselves alive. For example, as the anthropologist Marshall Sahlins has pointed out, traditional hunter-gatherer peoples generally only spent twelve to twenty hours a week searching for food, with the rest time spent in leisure activities like storytelling, singing and chatting.[14] It was only with the advent of agriculture – starting in the Middle East around 10,000 years ago – that survival started to be difficult. The diet of traditional hunter-gatherer peoples – apart from the small amount of meat they ate – was very similar to that of modern-day vegans (no dairy products and a wide variety of fruits, vegetables, roots and nuts, all eaten raw) and was probably healthier than that of most people today. In addition, hunter-gatherers were much less prone to illness than later peoples.

Many of the diseases which we're now susceptible to only arrived when we domesticated animals and started living close to them – for example, pigs and ducks passed the flu on, horses gave us colds, cows gave us the pox and dogs gave us the measles.[15] It's usual to think of the advent of agriculture as progress, but in actual fact it led to much more labour, a poorer diet, more illness and a shortening of lifespans.

Our ancestors in the Middle East and central Asia were some of the first groups to take up agriculture, and so life was already more difficult for them. And it became much more difficult at around 4000 BC, when a process of severe desiccation began to turn vast areas of the Middle East and Asia to desert, drying up water sources and making the land infertile. It may be, therefore, that our ancestors developed this desensitising mechanism in response to these difficulties, whereas other peoples may not have needed to.

This desensitising mechanism affects our perceptions of everything. The rule seems to be that when we're exposed to a new environment or experience, for a short time we're able to perceive its realness. But at a certain point the mechanism begins to act on it, editing out its realness so that we no longer give our attention to it. It becomes just another familiar thing which we see no reason to look at, like the houses on the other side of the street, the buildings we pass on the way to work every day, the faces of the people we work with, and even the sky and the moon and the stars above us. As Alison Gopnik puts it, comparing our automatic perception to the intense conscious perception of children: 'We literally don't see the familiar houses and streets on the well-worn route home, although of course, in some functional sense we must be visually taking them in.'[16]

The important point is that this mechanism is 'hard-wired' into us, so that we develop it naturally as we grow into adulthood. This is the repression that Ernest Becker talks about. We lose the intense vision of the world we have as children because, as we move into adulthood, the desensitising mechanism develops. In young children it probably isn't active at all, so that the world is always full of strangeness and wonder. They *don't* get used to the streets or buildings of their towns or to trees or to people's faces; no matter how many times they see or experience these things, their reality never fades. And as a result time passes very slowly for them.

Although the mechanism itself is fully developed by the time we reach adulthood, it causes a process of familiarisation which gradually intensifies all through our lives. As we've seen – and as William James noted – the longer we live, the *more* familiar the world becomes, and the less perceptual information we absorb. Our perceptions become progressively less fresh; a larger and larger proportion of them become filtered through this desensitising mechanism. And as the world becomes more familiar, we take in progressively less information from it, so that time gradually speeds up. Eventually the grey, shadowy half-reality of the world as seen through a filter of familiarity becomes our normal vision, and we come to assume that this is the correct and objective way of seeing the world.

New and old experience

There are two different ways in which our perceptions become progressively less fresh and more familiar. On the one hand – as William James and John Wearden point out – as we get older there's progressively less newness in our lives. The life of a twenty-year-old woman is still full of new experiences. She's still discovering new kinds of music, food, literature and other new hobbies and interests. She might be experiencing her first serious romantic relationship, learning to drive, going abroad for the first time, discovering new towns or the countryside close to where she lives and so on. When she has these new experiences she's free of the desensitising mechanism; she perceives the 'raw experience' of the world and processes a large amount of perceptual information.

The same person at the age of 30 might still be having a few new experiences. She might be having a baby, going abroad to a country she's never been to before, learning a new language, or starting a new job. But by the time she reaches 40 the world contains much less unfamiliarity. Her life probably consists mainly of the repetition of experiences which she's had hundreds or thousands of times before. She works at the job she's had for the last twenty years, goes home to the house she's lived in for the last ten, devotes her free time to the same hobbies and interests she discovered when she was twenty, goes away at weekends to the same countryside, to the same foreign country every year, and so on. Because of this repetition, the desensitising mechanism has a greater hold over her. She's hardly ever free of it, which means that she absorbs much less perceptual information.[17]

But if that was the *only* reason why our perceptions become less fresh – and why time speeds up – as we get older, there wouldn't be much difference between the time perceptions of a 40- and a 60-year-old person. Most of us use up almost all of our 'stock' of new experience by the time we reach middle-age, and so there'd be no real reason why time should appear to move faster for a person at these different ages. However, the second reason why our perceptions become less fresh is because as we get older, our repeated pattern of experiences becomes *more* familiar to us and progressively less real. It may be that every time you look at the buildings and streets of your hometown, every time you revisit the same shops and pubs, every time you do your day's work or even take your dog for a walk – the desensitising mechanism filters out a little more of the reality of these experiences, so that you give slightly less attention to them and absorb less information from them. In William James' words, 'each passing year converts some of this experience into automatic routine'.[18] As well as experiencing lots of new things, a woman at the age of twenty is still quite 'fresh' to the phenomenal world around her – but over the next twenty years, she'll look at the same street scenes and the same sky and the same trees hundreds of thousands of times, so that more and more of their realness will fade away.

It's important to point out though, that this doesn't *have* to be the case. To some extent it depends on the way that we choose to live our lives – and also on how we choose to respond to our experiences. In fact you could say that many of us aid this process of familiarisation – or at least don't do anything to fight it – by being content to keep our lives to a fairly narrow range of experience. If we so desired, we could make a conscious effort to expose ourselves to unfamiliarity – by travelling to countries we've never visited, by finding new hobbies and interests, or by changing our jobs every so often. In other words, we could make use of the second law of psychological time ('Time slows down when we are exposed to new experiences and environments') to stop time speeding up, or at least to slow down the speed of its increase. (We'll look at extending life by consciously exposing ourselves to unfamiliar experiences in more detail in the penultimate chapter.)

Similarly, there might be certain events in a person's life which bring in a flood of new and unfamiliar experiences, and so interrupt the increasing speed of time. These might be major life events such as war, divorce or bereavement. When I recently met the physicist Julian Barbour (author of *The End of Time*) he told me that he didn't feel that time was getting faster as he got older, because since the publication of his book a few years ago so many new things had happened to him, and his life was fuller than ever before. A lady on one of my courses recently told me that she didn't feel time was getting faster because since she'd retired two years ago she'd taken the opportunity to go on so many different courses and learn about so many new things. As I noted in the last chapter, I don't feel that time has sped up for me over the last three years, because of all the new experiences in my life since the birth of my first child. It's probable that time will start speeding up for me again in the near future, especially when both children are at school. I've often heard parents say things like: 'I can't believe how quickly my children have grown up – it just seems like yesterday when they were toddlers', or 'Enjoy it while you can, because before you know it they'll be teenagers', and I'll no doubt experience this too.

On the other hand, there are some people who don't seem as affected by the desensitising mechanism as others, and respond to their experience in a more real way. They see the world with something of the fresh, first-time vision of children all through their lives. These are the kind of people – sometimes seen as eccentrics by those around them – who often begin sentences with phrases like 'Isn't it strange that …?' or 'Have you ever wondered …?' They're the kind of people who might stop in the street to gaze up at a beautiful scene of the sun breaking through clouds or a silver moon above the rooftops; or they might stare intently at the sea, at flowers or at animals, as if they've never seen them before. As Alison Gopnik notes, poets and artists often have this kind of 'childlike' vision – in fact it's this that usually provides the inspiration for their work. They have a sense of strangeness and wonder about things that most of us take for granted, and feel a need to capture and frame their more intense perceptions. And this is no doubt true of some scientists too, who may be free from what Richard Dawkins describes as 'the anaesthetic of familiarity',[19] so struck by the strangeness and wonder of phenomena which most people take for granted that they feel a desire to understand and explain them. These people will be less affected by the first law of psychological time than others; time may well speed up for them, but perhaps not to the same degree.

The second law explained

I should reinforce that this Information Theory isn't the only factor in the speeding-up of time we experience. We've noted that there's another reason why time passes so slowly for children, besides the fact that they take in more perceptual information than adults: their less developed ego. Before the ego develops, babies seem to be outside time, and as it develops they gradually fall into a world of sequence and duration. And I'm also open to the possibility that the proportional and biological theories we've discussed that may add to the effect of increasing familiarisation – that every year makes up a smaller percentage of our life as a whole, and our metabolism slows down and our body temperature falls as we grow older. However, one reason why I believe that the Information Theory is the main factor is because it explains the other time distortions we experience.

To summarise, then, following James' and Ornstein's theories, one of the two concepts at the centre of my Unified Theory of Time is that: 'The speed of time is relative to the amount of information we absorb and process.' Or in the words of another researcher who links time perception to information, the Swedish psychologist Marianne Frankenhauser:

> We may assume that the experience of a certain duration is related to the total amount of experience (sensations, perceptions, cognitive and emotional processes, etc.) which takes place within this time period, in short, the *amount of mental content*.[20]

The more information we process – or the more mental content there is – the slower time goes. The less we process, the faster it goes. In other words, how fast or slow time seems to pass for us depends to large extent on how much we attend to the world around us, and how much of its reality we take in. The relationship between information-processing and states of absorption might seem confusing. In states of absorption – such as when you're listening to an interesting lecture or reading a good book – there's usually some input of information. But whereas absorption makes time pass quickly, processing information usually stretches time. So how do these two effects on time interact with each other? We'll deal with this question in the next chapter, when we look at exactly why time speeds up in states of absorption, and slows down in non-absorption.

Psychologists have suggested that the human nervous system is able to process up to 126 'bits' of information a second, or about half a million an hour.[21] In these terms, we could say that, on average, children – because they're so 'awake' to the world around them – process a very high quantity of information bits per hour. The purpose of the desensitising mechanism is to reduce the number of information bits which get through to us, since processing them takes up our attention and mental energy which could be used elsewhere. And so, as we grow older and the desensitising mechanism turns more of our experience to familiarity, our hourly average quantity of information bits progressively decreases.

This concept that processing information expands time explains the second law of psychological time ('Time slows down when we are exposed to new experiences and environments'). When we go to a foreign country, for example, we effectively become children again. You're surrounded by newness and strangeness, and it's all so *real* to you that you couldn't keep your attention fixed to your daydreams or distractions even if you wanted to. You perceive the raw experience of the world again – the strange buildings, the strange language, the strange fragrances of food and spices floating through the air, the new hotness and stickiness of the air, the unfamiliar traffic sounds, and so on. You stop thinking and doing (at least to a much larger extent than at home), and experience *being in the world*. As James notes, 'rapid and interesting travel' results in the same 'vivid apprehension' as childhood.[22] Or as Alison Gopnik puts it: 'As adults, when we are faced with the unfamiliar, when we fall in love with someone new, or when we travel to a new place, our consciousness of what is around us and inside us suddenly becomes more vivid and intense.'[23]

We've already seen that this applies to individual experiences too. When you undergo *any* new experience – the first time you fly, your first sexual relationship, the first months at your job, the first time you drive after passing your test, etc. – it's always much more real and intense than it becomes later, when the desensitising mechanism has acted on it.

Compare travelling on a bus for half an hour through the streets of your hometown with a bus ride through the streets of a town in India. In the former case, you pay scarcely any attention to your surroundings, just the bare amount needed to check the bus is going in the right direction and that you get off at the right stop. But in India your perception isn't automatic. You hungrily devour the reality of the situation, continually stare at the people around you and the buildings and street scenes you pass – and as a result you absorb a massive amount of information and the same half hour passes very slowly.

3

Absorption and Time: The Third and Fourth Laws Explained

A few years ago a colleague of mine had an operation to remove a cancer. He was under a general anaesthetic for several hours, but when he came round he didn't believe he'd actually had the operation, because no time seemed to have passed. 'All those hours just seemed to be wiped out', he says. 'One second I was talking to the anaesthetist and being given the anaesthetic and the next I was awake.'

This is a common experience with operations – the time between 'going under' and 'coming round' often seems to disappear. In a similar way, a woman who suffers from epilepsy remarks that when she has seizures, 'time does not pass. I come back from the seizure at the same instant that I went into it, although time has passed for others. Rather like they all went on a detour and I kept to the main path, and my one step was a circular mile for them.'

People who wake up from comas sometimes experience this too. They may have been unconscious for years and yet only feel that they've woken up after a night's sleep. In his book *Awakenings*, Oliver Sacks describes how he was able to use the drug L-DOPA to awaken people who had been catatonic since the 1920s, after contracting a brain infection known as 'sleeping sickness'. Some of the patients did understand that time had passed while they were unconscious, even though they still felt as if they were young and living in the 1920s. But for others these decades appeared to have, in Sacks' words, 'no subjective duration whatever'. It was this, he speculated, which enabled some of them to 'pick up where they'd left off' so easily and resume their old activities and behaviours as if they'd only been asleep for a night. If they'd experienced the time passing, they'd surely have forgotten these old behaviour patterns. And rather than remembering events from 40 years ago as if they'd happened yesterday, these memories would surely have been much less vivid.[1]

Time often disappears in a similar way when we're asleep. When we wake up in the morning we always have a sense that some time has passed, of course. And sometimes, after a short sleep – especially if we've had vivid dreams – we might feel that much more time has passed than actually has. But more commonly we have exactly the opposite experience. The alarm wakes you up in the morning and you press the button down and nod off again for what seems like a few moments – but then you wake up with a jolt and realise that half an hour has gone by and you're late for work. A short nap usually turns out to be longer than you think. After one of my lectures recently a woman stayed behind to tell me that she was a teacher in a prison, and that my description of how time disappears under anaesthetic reminded her of how prisoners try to 'sleep away' their sentences. The prisoners she worked with spent as much time as they could asleep, up to twelve hours a day, to try to make the time pass faster. And from a theoretical point of view this certainly would work. Assuming their experience of time is similar in other respects, a person who sleep twelve hours a night will live through less time than a person who only sleeps for seven hours – and over a period of several years this will add up to a very substantial difference.

In the case of short naps though, the important factor here is whether we dream or not. When we dream we do have a sense that some time has passed, because our consciousness is still absorbing and processing information. In fact sometimes it goes beyond this, to a deeper level, and time seems to expand massively in dreams. It's possible to live through several hours of time in a dream in just a few minutes of clock-time. (This time expansion is due to another factor – the dissolving of the ego in sleep – which we'll look at in the next chapter.) But when we *don't* dream – and scientists tell us that we only dream for 1.5–2 hours each night – time seems to vanish.

Something similar – although with a weaker effect – can happen when we watch television. As psychologist Marie Winn (author of *The Plug-in Drug*) has pointed out, television often has the function of a drug, which enables us to escape from reality.

> Not unlike drugs or alcohol, the TV experience allows the participant to blot out the real world and enter into a pleasurable and passive mental state. The worries and anxieties of reality are as effectively deferred by becoming absorbed in a TV programme as by going on a 'trip' induced by drugs or alcohol.[2]

In this respect watching television isn't so different from sleeping. Often when we watch TV (although this depends on what kind of programmes we're watching, of course) we aren't really awake or alive; we experience an intense state of absorption which is almost a suspended animation, similar to hypnosis. We forget ourselves and our surroundings and slip outside time, so that hours pass by almost without us noticing.

It's as if when the 'real world' disappears to us, then time disappears as well. The important point is that when you're under anaesthetic, in a coma, asleep, or watching television, you aren't conscious of the phenomenal world around you. You don't give any attention to it, and so don't absorb any perceptual information from it. It's true that you absorb a *degree* of information from television (the images and sounds of the programmes plus the actual content of the programmes), but this is only a measly trickle (or a small number of bits of information) compared to the wealth of sensory impressions and experiences which the phenomenal world offers.

It's also significant that in these states there's usually very little happening *inside* our heads too. Mental activity – thoughts, images, memories, daydreams, dreams and so on – is a kind of information too, of course, and when a person is under anaesthetic, watching an absorbing TV programme, this activity ceases. So we don't process any information from this source either.

And this is the key to the third law: 'Time passes quickly in states of absorption'. It does so because in these moments we process very little perceptual – and cognitive – information. At a mild level of absorption (say, when you're reading a book that's a bit boring, or watching a TV programme you've seen before) you give a degree of attention to your surroundings and so do absorb some perceptual information, which means that time moves at a moderate speed. But at an intense degree of absorption your attention is so firmly fixed to its object that you *completely* lose touch with the phenomenal world. You absorb so little perceptual information that 'hours pass like minutes'.

Of course, there might be other kinds of information – besides perceptual – which you'll still absorb and process in these situations. If you're absorbed in a particular activity there's always some input of information, so shouldn't this counteract the time-quickening effect of your absorption? But even an interesting lecture or book will only bring a small input of information compared to, say, walking through a city for half an hour, or sitting on a bus watching the buildings and pedestrians pass by. When our attention is narrowed to a specific source we don't have myriad sensory impressions flying at us from all directions, and the low input of information we *do* receive hardly affects our sense of vanishing time.

On my courses I usually include a guided meditation session. I do this partly because I believe that – as we'll see later – meditation is a way of expanding and even transcending time, but also because I'm curious to see how meditation affects my students' sense of time. And I almost always find that the students underestimate how much time has passed. For a meditation of around 15 minutes, most people estimate between 7 and 11 minutes. This makes sense in terms of the third law of psychological time: the students close themselves off to perceptual information by being still and shutting their eyes, and they concentrate on following the instructions I give them, so there's very little thought information running through their minds. In other words, they are in a fairly intense state of absorption.

Sometimes people fall asleep in the meditation, and they usually estimate even less time – typically, between 3 and 5 minutes. (Occasionally there is an opposite effect and a person *over*estimates how much time has passed. For example, on my most recent course we meditated for 13 minutes, and there were two ladies who estimated 16 and 20 minutes respectively, while the average estimate of the rest of the group was around 7 minutes. There's a reason for this, however, which we'll look at in more detail later on. In *deep* states of meditation we transcend our normal ego and in the process, seem to transcend time. And it is, I believe, very significant that these two were the only members of the group who'd done meditation before, with the lady who estimated 20 minutes the only person who meditated regularly.)

A different kind of busyness

'Busy time' doesn't *always* pass quickly though. It depends on the kind of busyness. There's a kind which doesn't allow a state of absorption – at least in the normal sense – to develop, and which therefore doesn't have a 'time-shortening' effect.

The sociologist Gary Alan Fine made a study of how restaurant chefs perceived time, and found that they regularly experienced three different kinds of 'time distortions'. When there weren't many customers and they weren't very busy, time seemed to pass slowly (the fourth law of psychological time). But when they were busy they could have two different experiences of time: when they were busy and things ran smoothly, they found themselves getting to closing time and wondering where all the time had gone (the third law). But when they were busy – usually their busiest of all – and things didn't run smoothly, when there were too many demands on them and too many things to think of at the same time, they found that time went slowly.[3]

One of my students gave me a similar description. He had part-time job selling programmes at football matches. Most of the time he wasn't busy and time seemed to pass slowly, but at very intense periods this changed radically.

I had to sell the programme for 3 hours before the match. For the first 2 hours it was always very slow. Not many people came up and most of the time I was just staring into space and daydreaming, feeling bored. Time went very slowly and I used to ask myself how I was ever going to get through the time that was left to kick off.

But about 45 minutes before the kick-off it always went mad. People crowded around me and in the last half hour I could sell anything up to 300 programmes. On this particular day, it was even busier than usual. There were sometimes a dozen people around me at the same time, all thrusting their money at me in different kinds of coins and notes. Selling a programme sounds like something a monkey could do but when it was busy like this it was actually quite skilled. I had to concentrate hard and do some quite complicated things at the same time and at a very high speed. I had to take the different coins and notes and put them in the correct place in my money belt, take the correct change from the different parts of money belt and give it to the right person in the crowd, and then hand over a copy of the programme. I had to do all of that with one hand – with the other I had to hold up a copy of the programme, so that other people coming up could see it. And at the same time I had to carry on shouting the name of the programme to attract other customers.

On the day I'm talking about, there was one really intense period when I was working so fast that my hands seemed to be in a blur in front of me. At one point, between 2.30 pm and 2.35 pm, I must have sold around 50 copies. I remember looking at my watch and seeing it was still only 25-to and being amazed, thinking that at least 15 minutes had gone by. So much time seemed to have passed in those 5 minutes, because I was so busy.

This might seem to contradict the third law of psychological time – after all, this programme seller and the restaurant workers were extremely busy, their attention was focused on their jobs, so time should have gone quickly to them. But in these situations we aren't 'absorbed' in the same way that we normally are when we're busy. Absorption usually means a steady focus on one particular object, such as a TV programme, a game of chess, or a book. But in situations where we're dealing with a lot of different demands at the same time, and where unforeseen problems are occurring, this doesn't happen. Your attention continually flits around, from one object to the next, and you don't focus on one particular task for long enough to become absorbed in it. And it's the very fact that you're not absorbed which makes time pass slowly. As we've just seen, in states of absorption the flow of perceptual information from our environment falls to a very low level – but in these situations, we take in a high volume of information.

This is particularly clear from the programme seller's description of the massive amount of activity and information-processing he packed into those 5 minutes. And we can imagine the same thing occurring at a very busy time in the restaurant, with the chefs struggling to handle the volume of orders, cook several meals at the same time, and deal with problems like running out of ingredients or equipment not working properly.

The fourth law explained: thought-chatter

At first sight the fourth law ('Time passes slowly in states of non-absorption') may not seem to make much sense in 'perceptual' terms. But in states of boredom or non-absorption we aren't particularly responsive to our experience. Although we might be more conscious of our environment than in states of absorption, we certainly aren't alert and attentive to our experience in the same way that we are in unfamiliar situations. (In fact if we were, we wouldn't be bored in the first place.) But the important thing here is that when our minds aren't occupied they process a massive amount of information from another source.

Why is it that we find states of non-absorption so horrible, and do everything we can to avoid them? Why do we need to kill time by watching videos on long coach journeys or by reading magazines in the dentist's waiting room? Why do we always try to make sure that there are no empty spaces of time in our days, so that we spend large parts of our life doing things just for the sake of having something to do, or watching TV programmes that we're not particularly interested in?

The answer can be summarised in one word: thoughts. When our minds aren't occupied they chatter away to themselves. A chaotic stream of images, ideas, memories and other kinds of thought information flows through them at a lightning speed. As an experiment now, close your eyes and look inside your mind. After a few seconds you'll be aware of thoughts whizzing through your consciousness – thoughts about what you've got to do later today, fragments of songs you heard on the radio this morning, memories of what you did yesterday or last week, daydreams about things you'd like to happen, etc. It's as if there's a kind of film going on in your mind all the time – only one made by a mad director, which shifts from one scene to another ten times every second and is completely random and chaotic, with no plot. Let your mind carry on thinking, let it go wherever it wants to, and after a minute or so stop and try to retrace your thoughts, all the different ones which led you to this particular thought in your mind now. Most likely you'll be amazed by the different leaps your thoughts have made, the different tangents they've gone off on, and how much ground they've covered in just a minute.

As an experiment to test his memory, the 19th-century English psychologist Francis Galton wrote down a series of words (such as 'carriage', 'abbey' and 'afternoon') and paid attention to the thoughts and memories that they aroused as he read through them. He did this on four separate occasions, and finished with a total of 505 associations in 660 seconds. This is over 50 individual thoughts or memories per minute, just over one per second, but Galton felt that it was still 'miserably slow' compared to the speed at which his thoughts worked under normal circumstances.[4]

A few days before writing this I was sitting with my wife, looking out at our back garden. A squirrel ran by and for some reason – it seems nonsensical now, but I'm sure it related to something – my wife said: 'What's the sound of one squirrel clapping?' This set me off on a train of thought, and after about 15 seconds my wife noticed my abstracted look and asked: 'What are you thinking about?' I stopped and realised that I was thinking about Tony Blair, and the fact that he'd recently won the general election. I was amazed that I'd got from 'What's the sound of one squirrel clapping?' to Tony Blair in seconds, and made an attempt to trace my thoughts back. They went something like this: my wife's question reminded me of one of my favourite songs by Van Morrison, 'Enlightenment', in which he sings: 'What's the sound of one hand clapping?' I sang a couple of lines in my head, and remembered hearing someone singing the song at the yoga centre I went to. I remembered that I always thought that was strange because the song's not actually about being enlightened, but about being confused about what enlightenment is. Another couple of lines from the song ran through my head: 'Every second, every minute, it keeps changing to something different. Enlightenment, don't know what it is.' Then I remembered the brief interview I'd heard with Van Morrison on the radio a few days ago. The interviewer said that he'd been told not to bring up the subject of President Bush and his iPod. Apparently the Whitehouse recently released details of the contents of the President's iPod, and Van's old song 'Brown Eyed Girl' was one of the songs on it. And naturally enough, George Bush led me on to Tony Blair and the general election...

And all of that – and probably more details that I've forgotten – in about 15 seconds. It's no wonder that, in his novel *Ulysses*, it takes James Joyce more than 50 pages to describe the thought-chatter running through his character Molly Bloom's head as she's lying in bed waiting to sleep.

As another experiment, choose an object in your room – say, a picture on your wall or a plant on your window ledge – and concentrate on it. Try to see everything about it, examine it in as much detail as you can, and keep your attention focused on it for as long as possible ...

What you'll probably find is that you can concentrate on it only for a short while, perhaps even just a few seconds, because the thought-chatter whizzing through your mind gets hold of your attention instead. You find that instead of focusing on the single object, you're on the other side of the world, thinking about your friend in Australia, or ten years in the past, remembering that there was a plant like this in your parent's house and recalling all kind of scenes from the time when you lived there.

We experience this thought-chatter whenever our attention isn't absorbed in external things. It usually isn't there when we watch TV or read magazines, because our attention is too focused to allow it into our minds. But it's usually there when our attention isn't absorbed: when we do jobs which aren't very interesting or demanding, or when you're travelling to work on the bus or train without a book or paper to read. It's also there when you're on your own in an empty bar waiting for a friend to turn up, and when you lie in bed at night trying to get to sleep. In the latter situation there's nothing external to occupy your attention at all; you're left completely alone with your own mind, and often your thought-chatter becomes so powerful that it stops you getting to sleep. Some people find their minds 'race' so much that they take sleeping tablets to try to slow it down.

As the studies of Mihaly Csikszentmihalyi have shown, when this thought-chatter continues for any length of time, it's usually a negative experience. We experience a state of what he calls psychic entropy, in which we have no control over our minds.[5] We also feel a discomforting sense of isolation and separation, since we're alone inside our heads with these thoughts, while the entire world and everybody else are 'out there', on the other side of our skulls. Our thought-chatter is usually negatively based as well, centred around worries and problems, and so often triggers feelings of anxiety and depression, or bitterness and jealousy.

But most importantly – as far as we're concerned – when we experience this thought-chatter, time passes slowly to us.

There are some types of thought-chatter which absorb your attention and make time pass quickly – daydreaming, for example, which takes you out of yourself in a similar way to television. But most thought-chatter has the opposite effect: the massive amount of information it contains stretches time in the same way that perceptual information does.

This is the key to the fourth law of psychological time, explaining why a 'watched pot never boils'. If you distract yourself while waiting for the pot to boil, it does so quite quickly (although the danger here is that it might boil over because you're not paying proper attention). But if you just stand there watching it, a massive amount of thought information flows through your unoccupied mind. All of this information stretches time, making you feel that you've been waiting for longer than you have. And the same with a dreary lecture or boring sports match – in these moments your attention isn't absorbed by external stimuli and as a result your consciousness processes a large amount of thought-chatter.[6]

This helps to explain another paradox of time: that 'empty' time might seem to drag when we actually experience it, when we look back at it from the future it seems constricted, and may feel shorter than it was. As William James notes, for example, a month-long period of illness might seem interminable at the time, but in recollection it shrinks into nothing, because it 'hardly yields more memories than a day'.[7] We've seen that empty time drags because of the massive amount of 'thought information' we process when our minds are unoccupied. But in the future, when we look back at these empty periods, what Robert Ornstein refers to as 'information storage' becomes important. Periods of boredom and inactivity leave us with much fewer memories than active times. We don't 'store' many memories from them, simply because so little happens to us. As a result, from a future perspective, these periods are hardly 'stretched' with information at all.

Pain and distress

Have you ever noticed how an hour visit to the dentist can seem to take up the whole afternoon? This isn't just because of the long half hour you spend in a state of nervous non-absorption in the waiting room, but also because of the pain the dentist (depending on how good he/she is at the job) inflicts on you during the treatment. As one man told me:

> Every time I go out of the door at the dentist's I feel like it was hours since I went in. Things seem different, as if I've been away for a long time. I'm usually only in there for half an hour but when he's drilling my teeth I'm aware of every tiny moment passing and it seems to last for ages.

In a similar way, here a woman describes the painful experience of having an abortion without anaesthetic:

> I bit my lip and clenched my fists; fortunately I have no fingernails, or I'm sure they would have gone right through my hands. It seemed like it would never end and it was *terribly* painful. I suppose it took ten minutes, but it felt like two hours.[8]

As these reports show, physical pain can make time pass much more slowly than normal. The same is true of psychological and emotional types of pain and distress, such as depression, fear and anxiety. Any kind of suffering stretches time. A young man who was depressed and upset because of problems with his girlfriend, for example, describes how he 'went out to the sea wall to think about things. I was feeling very ... isolated and angry. Time seemed very slow.'[9] As Dr Larry Dossey summarises: 'Minutes seem like hours when one is hurting.'[10] A friend of mine who suffers from depression, anxiety and panic attacks told me how psychological pain alters his experience of time.

> When it's unbearable the pain makes time much longer. The more you're in pain, the more time there is. Pain is something that doesn't have time. It's infinite. It starts and I have the impression that there's no time ...
>
> Periods of time are no longer there. When I have panic attacks I completely lose awareness of time. I don't know whether it's 10 minutes, an hour or a whole day.

These effects have been verified by a number of studies. A study of air travel by the Israeli psychologist Dinah Avni-Babad showed that people who fly regularly 'experience a swifter passage of time' than people who don't.[11] The reason for this, Avni-Babad suggests, is that the people who don't fly often feel less secure and comfortable, and their anxiety stretches time. Researchers at the University of Barcelona made a study of a large group of seriously ill people, asking them the question 'How long did yesterday seem to you?' at various stages, as their health and their moods fluctuated. They found that there was a strong correlation between how bad the people felt at that particular time, and how long they felt the day lasted. The worse they felt, the more time there seemed to be in a day.[12] And a study of people suffering from depression (by the psychiatrists Wyrick and Wyrick) found that they frequently overestimated how much time had gone by, and complained that time was moving slowly to them.[13]

It's quite tempting to think of this as a law of psychological time in itself – 'Time passes slowly in states of discomfort and distress'. But in a sense this effect is covered by the fourth law. One of the worst things about pain or distress is that it's so difficult to divert our attention away from it. The sensation becomes the focus of our attention, and resists our attempts to absorb our minds with other things. It partly depends on *how* powerful the sensation is, of course. If you're suffering from toothache it might be possible to take your mind off the pain by reading a book or watching TV, although the pain will always return the moment you relax your attention. But if you're suffering from acute mental distress and/or severe physical problems, even this may not be possible – the pain and distress will override every other stimulus you offer your mind. As I found on my long-haul flight from Manchester to Singapore, the reason why anxiety makes time go slowly is simply because it refuses to let you fall asleep, concentrate on a film, read a book or even talk carefreely with the person sitting next to you. The anxiety fixes your attention to the moment-to-moment reality of the situation.

'Time-consciousness' is probably a factor here, especially if you know that your distress only has a limited duration, and will end some point soon. You'll be aware of every moment passing and desperately anticipate the end of the period – the end of the flight or the end of the drilling at the dentist's – and this acute awareness will magnify time.

But probably the most important factor is that when you experience pain or distress you're likely to be in a state of non-absorption – in which a large amount of thought-information will pass through your mind, and so slow down time. There's some evidence for this in a study carried out by psychologists at Penn State University, in which twenty abstaining smokers were asked to estimate a 45-second time interval. Whereas non-smokers were more or less correct in their estimates, after 24 hours without a cigarette the smokers all overestimated the interval by large amounts – from 25 to 60 seconds. And as one of the directors of the study, Dr Laura Klein, noted, it was significant that the abstaining smokers 'reported feeling more stressed and unable to focus or be attentive'. This led her to conclude that: 'Time estimation [can be] used as an index of attention processes.'[14]

Hurtling towards death

So far in this book we've been looking at time distortions, the moments when time seems to pass at a different speed to usual. But what about our *normal* sense of time passing? What is the 'normal' from which these time distortions deviate?[15]

Some scientists believe that our sense of time passing comes from a kind of 'pacemaker' in our brains. This produces regular pulses which are counted up by an 'accumulator' and stored in our memory. At the end of a period of time we know how many times the pacemaker has pulsed, and so have a rough idea of have much time has passed – it's a bit like a having a clock inside your head.[16] As yet there's no clear evidence for this theory, but it does seem logical to a degree – it would explain how many of us are able to estimate the time fairly accurately, even if we haven't seen a clock for a long period. However, since the time distortions we experience can be so extreme, this pacemaker would have to be very flexible and be very easily influenced by psychological states. It would have to speed up in states of absorption and slow down with new experiences, and stop altogether in states of egoless-ness. I believe it's unlikely that any brain mechanism could function in this way, and that it's much more probable that these different time perceptions are caused by different levels of information.

In psychological terms, our normal sense of time is simply when we *don't* experience any time distortions. It's when we aren't in a state of intense absorption or of boredom, and when we aren't exposed to unfamiliar environments or experiences. In other words, it's when we're in what's a normal state of consciousness for us.

The disturbing thing about this, though, is that the 'normal' state of consciousness for us is one in which time passes very quickly to begin with. Of the four laws of psychological time, it's certainly the third ('Time passes quickly in states of absorption') which we experience the most. The first law ('Time speeds up as we get older') is always acting in our lives, but it's so gradual that we don't often notice its presence; we don't expose ourselves to new experiences and environments frequently enough to experience the second law so much; and we try to avoid situations in which we're unabsorbed and experience the fourth law, since we usually find them unpleasant. On the other hand, the third law is almost always operating in a very obvious way in our lives. As we've already noted, most of us spend a large part of our days with our attention absorbed in external things to a greater or lesser degree. We distract ourselves with radio or newspapers while we have breakfast or travel to work, and we spend most of the next eight or nine hours with our attention focused on the tasks of our jobs (providing they're interesting and challenging enough). And then we spend our evenings watching TV or with our attention absorbed in other distractions. At weekends we have to work a little harder to keep our attention occupied: we give ourselves jobs to do around the house, clean the car or do the garden, go to see friends, to restaurants or the cinema and so on.

Our normal state of being is, therefore, one in which our attention is absorbed in external things. And because of this, time is *always* going quickly to us without us realising it. Our normal sense of time is one in which it's already passing quickly. (Which is why, as we've noted, we aren't actually aware that this third law is operating until we become *intensely* absorbed in something.)

You could say that there's something inherently nihilistic about this. What's the point of being alive if we spend so much of our time in a state of *absence* from the world? We already have to sleep for eight hours a night, so why should we add to this by spending our daytime hours – particularly the evenings we spend watching TV – in a similar state of suspended animation? (This admittedly doesn't apply to states of active absorption or flow though – as Csikszentmihalyi has shown, flow is a positive and healthy experience which gives rise to happiness and helps form a dynamic and well-integrated personality.)

But simply in terms of time, this way of life doesn't make any sense at all. We all want to have as much time in our lives as possible, and yet we let so much be eaten away by distractions. Time is already speeding up for us, since it's getting faster as we get older, and we actually *add* to this acceleration by constantly subjecting ourselves to the third law of psychological time.

In a similar way, we let time speed up for us by *not* taking advantage of the second law of psychological time ('Time slows down when we are exposed to new experiences and environments'). Many of us are content to stay in the same life-situations for years on end, living in the same home territory and repeating the same experiences. We rarely expose ourselves to different places or activities, and so the desensitising mechanism takes greater and greater hold over our perceptions of the world.

All of this makes nonsense of our attempts to live for as long as possible. Why strive so hard to add more time to our lives by eating healthy food, by exercising, or even by having our bodies frozen when we die, when we actually take so much time away from ourselves?

However, what's our alternative? The problem is that states of absorption are so pleasant, while boredom and non-absorption are so *un*pleasant. We can't help wanting to spend as much time as possible in the former, even though they make time pass so quickly. If it's a choice between spending the evening watching television or spending it mulling over the problems in our lives or letting chaotic thoughts whiz through our minds, then we're obviously going to choose the former.

But there is a way out of this trap. As we'll see later, we don't have to choose between a pleasant state of absorption and an unpleasant state of non-absorption. We actually have another choice, a third state of being which avoids the problems of these first two.

Before we look at this though, we have to try to complete a Unified Theory of Time by examining the second major factor – besides information – which affects our personal sense of time.

4

When Time Stands Still: The Fifth Law of Psychological Time

A few years ago a friend of mine was knocked off his motorbike at a high speed. The impact catapulted him through the air, and when he hit the ground he lost consciousness. The next thing he knew he woke up in hospital eight hours later. But he can still vividly remember the few moments before he lost consciousness.

> I saw the car's windscreen shattering. The glass sprayed out so slowly, like a fan, and it looked beautiful. All the pieces were shining in the sun. I felt like I was floating through the air, almost as if I wasn't going to come down. I looked into the sun and it was like being on a plane, when you're above the clouds and it's a brilliant white colour.

Accidents and other moments of sudden shock often bring an extreme slowing down of time. Here a person who regularly goes riding told me how she experiences time when she falls off her horse.

I am always surprised when I fall off a horse by how long it takes me to hit the ground! I feel as if I am floating through the air and I always think that I am going to land really softly so I relax and wait to land. And then, when I hit the ground and am lying there winded, I am completely surprised.

The same stretching of time often occurs in emergency situations, when our lives are threatened. There are many reports of time slowing down massively during earthquakes, for example. One woman who experienced the devastating earthquake in Armenia in 1988 reported that: 'It was like a slow-motion movie. There was a concrete panel slowly falling down.'[1] Similarly, the photographer Ansel Adams describes being woken up by an earthquake at 5 am, with his bed crashing against the walls and windows smashing. 'The roaring, swaying, moving, and grinding continued for what seemed like a long time', he writes. 'It actually took less than a minute.'[2]

In *The Dance of Life*, Edward T. Hall reports the case of a Navy test pilot who had taken off from his aircraft carrier only to realise that his plane wasn't powering up. 'Everything went into slow motion', the pilot told an interviewer. 'After about one second, about seventy-five feet after I started rolling, I knew I was in deep trouble.' What he did over the next eight seconds in order to save himself was so complex and detailed that it took him over 45 minutes to describe. Hall notes that he had a similar experience himself, when a mountain lion escaped from its cage, and he was trapped in a narrow passageway with it.

> My first awareness of what had happened was when I felt something brush by the calf of my right leg. Then as I watched the lion lick a spot of grease next to my toe, time slowed down ... Putting years of experience with animals to work, while I mentally reviewed and rejected a half dozen options and their scenarios, the only workable solution seemed to be to make friends.[3]

In the Zone

Many professional sportspeople are familiar with these kinds of experiences. This might seem anomalous at first, until we consider that sport often involves an artificial recreation of emergency situations. It doesn't *actually* matter very much whether a golfer wins a tournament or a goalkeeper makes a save in the last minute of a match, but the players and the crowd treat these situations as if they are a matter of life and death. And sometimes sportspeople find themselves entering the Zone, a state in which everything 'clicks' and they shift to a higher level of performance. New reserves of effort and energy are released inside them and they suddenly become capable of feats which would normally be beyond them. Everything they do seems naturally and inevitably perfect, without any extra effort. Time moves much more slowly than normal – in fact, this is often the main reason *why* players are capable of such astounding feats; they have more time to play with, more time to anticipate their opponents' actions and to position themselves. The American tennis champion Jimmy Connors, for example, describes how when he was in the Zone the ball seemed to grow in size as it came over the net, and moved so slowly that he had all the time in the world to choose how and when to hit it.[4]

Similarly, the New Age writer and speaker David Icke was once a professional goalkeeper, and found that: 'All my best saves were in slow motion.' He recalls how once, when he was playing in an important FA Cup match, somebody fired a shot from close range that looked unstoppable.

In conscious time, it was soaring for the top left hand corner. No chance whatever for the keeper …

But – and I still recall this so clearly – as the Barnet guy made contact everything went into slow motion for me. I moved across, watching the ball drifting slowly to my left and then I dived, lifting my right hand to turn it over the bar. All was like a slow-mo replay and everything was quiet, like some mystical dream, until my right hand made contact with the ball. Then everything zipped back into conscious time, I landed and bounced on the floor and the noise erupted, as if someone had turned off the mute button.[5]

The American footballer, John Brodie experienced this slowing down of time regularly. As he describes it:

Time seems to slow down in an uncanny way, as if everything were moving in slow motion. It seems as if I have all the time in the world to watch the receivers run their patterns and yet I know the defensive line is coming at me just as fast as ever. I know perfectly well how hard and fast those guys are coming and yet the whole thing seems like a movie or dance in slow motion.[6]

It seems clear that these are essentially the same experiences as the accident and emergency situations I quoted above. It's significant that Brodie says that the experience only occurred at the most intense moments of games, when he was acting as if there were a genuine emergency.

Players of team sports don't always treat the game as an emergency situation – there are usually long lulls and periods when they aren't individually involved. But some short-distance athletic events are effectively an emergency situation for their whole duration, involving intense bursts of effort and concentration. It's not surprising, therefore, that many athletes report an extreme slowing down of time when they compete. According to the runner Steve Williams, for example: 'If you do a 100 right ... that 10 seconds seems like 60. Time switches to slow motion.'[7] While according to the swimmer Courtney Allen, when she does a freestyle race: 'Every detail seems to last forever. When you replay it in your head, it just goes on forever.'[8]

These experiences are very temporary, but it seems that the very *best* sportsmen are always in the Zone to a degree, or at least have the ability to slip into it regularly. This is also true of martial arts experts, who train themselves to slow down their perception of time, so that they see their opponent's movements in slow motion and always have enough time to react. According to his opponents, what made the Australian cricketer Don Bradman the best batsman who ever lived – with an international average twice as high as most other great batsmen – was the amazing amount of time he seemed to have to play his shots. Even though he never committed himself to a shot until the last moment, he always had more than ample time to position himself and find the correct stroke, as if the fraction of a second it takes a ball to reach a batsman from the arm of a fast bowler contained more time to him than for anybody else. The same is true of the great Irish footballer George Best. TV clips show him gliding effortlessly past clumsy defenders, who lurch at his legs to hack him down but are never quick enough. What's most striking is that he never gives the impression of hurrying himself – in fact he seems to be moving slowly as if, again, he's in a different 'time world' to his opponents. And in fact, Best himself said that he was always surprised at how slowly his opponents seemed to move, and at the amount of time he had to anticipate their tackles.

In other words, this may be the main factor that accounts for greatness in sport – a slower sense of time that gives the person more time for anticipation, reaction and movement. The ability to slip into the Zone at will could be much more valuable than any amount of fitness or technical expertise. And there are now, in fact, sports coaches who claim to be able to train sportspeople in this way. One of them, Mike Hall from Edinburgh, says he has become able to enter the Zone at will after practising t'ai chi for twelve years. He uses the ability when he plays squash, and describes it as 'a feeling of stillness, like I'm not trapped in sequential time any more. The ball still darts around, but it moves around the court at different speeds depending on the circumstances.' He believes it's possible for any athlete to learn this ability through mental training, so that they can, in his words, 'go faster by going slower'.[9]

Immobilising the ego

These experiences of time standing still don't seem to make much sense in terms of the Information Theory of time. It's true that during them we're likely to process a larger amount of perceptual information than normal (such as when my friend had his motorcycle accident), but it may be that this isn't the cause of the slowing down of time, but rather the effect of it. If time is moving slowly to you to begin with, then this allows you to absorb more perceptual information, since you have more time to see and hear what's happening around you.

Hall suggests that human beings have developed this ability to expand time at moments of crisis as a kind of survival aid. It gives us a much greater chance of surviving difficult situations simply because it gives us more time to think, to make decisions and try out different options. If the test pilot had been operating on normal time, for example, he would never have been able to think in such detail or to do so much, and probably wouldn't have survived. Again, this could apply to the Zone experiences of sportspeople too, since they usually occur in artificial life-or-death situations.

However, the real key to these experiences is the ego – or rather, the *absence* of the ego during them. The term 'ego' is often used to refer to the part of our personality which can be proud, selfish, vain or competitive – for example: 'He's got such a big ego; he thinks he's something really special'; or 'He hasn't got any ego at all – he doesn't care about credit or recognition'. But as I mentioned in the Introduction, in psychology ego usually refers to the 'I' in our heads, the conscious self which thinks and interprets the world and deliberates and makes decisions and plans. (This is essentially Freud's definition of the ego.) If you did as I suggested in the last chapter and closed your eyes and watched the thought-chatter flowing through your mind, you were observing the surface activity of your ego. In terms of its structure though, the ego is made up of a number of different components, such as your memories, your self-image (which might include your sense of being a success or a failure, or of being confident or inadequate), your beliefs about the world, and your sense of identity (belonging to a certain social or national group, a certain gender, age group, and so on). This is the 'ego structure' which creates your sense of being 'someone', an 'I' inside your head.

In Chapter 1 we saw that our sense of time passing increases as we move through childhood and our ego develops. During the first period of our lives, before the sense of self develops, we have no sense of time passing. (Likewise, in the next chapter, we'll see how some of the world's indigenous peoples were traditionally less subject to time than us, and that this is partly because of their weaker ego structure.) This suggests that if our normal ego was absent or in abeyance for some reason, there'd be no sense of time passing – or at least that if the ego-structure was weaker, our sense of time passing would be less pronounced, so that it would pass more slowly to us.

And this is exactly what happens in these experiences: for a short period, we lose our sense of being a thinking 'I' – or at least this becomes much weaker than normal – and so we lose our normal sense of time passing. Our conscious thinking mind is immobilised, and we cease to be subject to time.

It's important to point out that an egoless state isn't the same as a state of unconsciousness. A person who reaches an egoless state after meditation or an accident or during a sports match still has a sense of identity and is still conscious of what's happening to them. The point is that it's a *different* sense of identity. The normal ego structure – with its chattering thoughts and its beliefs and memories and sense of self-worth – fades away, but that doesn't leave a vacuum. There's still an *awareness* left. Think about what happens when a person is in a state of deep meditation, for example. Her mind might be completely still and empty, with no thoughts or other mental activity, but she still has an awareness of her experience; she's still aware of herself existing in her own mental space and of feelings of serenity and fullness inside.

In this sense, the philosopher Descartes was wrong when he famously stated: 'I think, therefore I am.' In some situations (such as meditation) you can stop thinking and *still* be, with an awareness inside you which exists independently of your thinking ego. Even in normal circumstances, when you're living day-to-day life and your mind is buzzing with thoughts, there's still an awareness which stands apart from your thoughts, and which enables you to be *aware* of them. We're sometimes aware of ourselves thinking thoughts that we don't want to think – about sex, for example, about people we don't like, or bad things that we don't want to happen (e.g. imagining that the plane you're on is going to crash, or that loved ones who are late have had an accident). In a sense we have two different identities: the thinking ego-self, and a deeper, more fundamental part of our minds that observes and witnesses reality (including our own thinking). This idea is an important part of the ancient Indian philosophy of Vedanta (as expressed in *The Upanishads*), which refers to the surface thinking mind as *jiva*, and the witnessing consciousness as *atman*. In moments of timelessness the ego-self (*jiva*) fades away, but our observing consciousness (*atman*) is still there.

It's also possible to think of this in terms of the two different hemispheres of the brain. Simplistically speaking, the left side of the brain is the 'scientist' in us, which thinks verbally and logically, while the right side is the 'artist', who is intuitive, creative and sees the world more holistically. Psychologists have suggested that the left side of the brain controls our sense of time, whereas the right brain is essentially 'time free'.[10] At any particular moment, we may be in a left brain or a right brain mode of thinking or perceiving the world, and it's when we're in the 'R-mode' that time seems to slow down or disappear. According to this view, then, when people face accidents and emergencies, or enter the Zone, they switch over fully into the 'R-mode'. Ultimately, though, we're still talking about ego-transcendence here, since the ego 'lives' in the left side of the brain. As the psychologist Brian Lancaster puts it, our normal sense of self is 'bound up with ... a language-based *interpreter* situated in the left hemisphere'.[11] Switching out of the L-mode and into the R-mode essentially means transcending the ego.

This relationship between time and the ego constitutes the fifth law of psychological time, that:

> Time often passes slowly, or stops altogether, in situations where the 'conscious mind' or normal ego is in abeyance.

Super-attention and shock

There are basically two ways in which our normal conscious mind can be immobilised. The first is as a result of intense concentration or attention, which is what happens in Zone experiences.

One of the apparent paradoxes of time perception is that while time goes quickly when our attention is focused on tasks or distractions, when our attention is focused to *a very intense degree* the opposite often happens: time slows down drastically. Here, for example, the author of *A Man on the Moon* records the reactions of the astronauts of the Apollo 8 mission, when they were waiting for the spacecraft's engine to fire up part way through their journey.

> Time seemed to slow down. Each man knew the engine must fire for the prescribed duration – no more, no less.

If the engine shut down prematurely, or it didn't deliver the proper amount of thrust, they could end up in a weird, errant orbit. If it fired for a few seconds too long, Apollo 8 would lose so much energy that it would crash into the moon. By the 2-minute mark the burn had begun to seem very long. Borman said aloud, 'Jesus, four minutes?'

'Longest four minutes I ever spent,' Lovell said as the engine roared silently into the vacuum.[12]

In an ordinary state of absorption these minutes would have passed quickly, but it seems that the extra level of attention that the astronauts were giving to the engine changed this. In a similar way, here a racing driver describes how time slows down during her races.

Each lap, a few cars sliding backward – the right direction. The experience of the last five years plus the shared knowledge of the guys who have been so helpful … Elastic time. Seconds become hours, and minutes an eternity.[13]

The key to this 'paradox' is that when our attention reaches an extremely intense degree, we cross over into a different realm, a new state of consciousness, in which we are affected by the fifth law of psychological time rather than the third. The intense attention immobilises our normal conscious mind, and we gain a new relationship to time.

A good example of the effect of this intense attention – super-attention, it might be called – is meditation. In meditation we focus our attention on a mantra (or perhaps our breathing or a candle flame) and if we can keep it focused, we begin to experience a state of mental quietness. Our thoughts begin to slow down, until eventually (if we're having a really successful meditation), they fade away altogether, along with our normal ego. We experience a state of egoless-ness, of no longer having an 'I' chattering away inside our heads. The act of concentrating our attention has the effect of dissolving the ego. As we'll see later, this is why states of deep meditation bring a sense of timelessness and can help us expand our sense of time in our daily lives.

The super-attention which sports involve can also immobilise the ego. A golfer or racing driver concentrates hard for hours, and eventually, when he or she reaches a certain intense pitch of concentration, their normal ego may fade away – and as a consequence, so may their sense of time. A short-distance athlete doesn't need to keep her mind focused for as long during the actual event, of course, but will aim to build up a state of intense 'focusedness' for hours beforehand, ridding her mind of other thoughts and distractions.

In view of this, it's not surprising that the greatest sportsmen are often famous for their incredible powers of concentration. There's an obvious link: as I've suggested, greatness in sport may largely be the result of the slowed down sense of time that occurs in the Zone, which you can only enter in states of intense concentration. The tennis player Billie Jean King, for example, was so focused that she was unaware of the crowd and even her opponent. All she was aware of was the ball and the face of her racket. 'It's like I'm out there by myself', she said. 'I appreciate what my opponent is doing, but in a detached, abstract way, like an observer in the next room.'[14] In a similar way, the British golfer Tony Jacklin wrote that he played his best games in 'a cocoon of concentration' in which he was 'living *fully* in the present, not moving out of it … I'm absolutely engaged, involved in what I'm doing at that particular moment.'[15]

The second way in which the ego can also be immobilised is by sudden shocks. Teachers of Zen Buddhism try to catch the egos of students off guard by hitting them with sticks while they're meditating, by giving them nonsensical answers to questions or by giving them *koans* (or riddles) to solve. The student can puzzle over *koans* for hours before his or her ego is finally immobilised by strain and confusion. It's taken for granted that when the ego is immobilised an experience of *satori* (or a higher state of consciousness) will result. And this is probably what happens in accidents and emergencies. The sheer shock of the car crash, the engine failure, the earthquake or the violent encounter, paralyses the ego. The events happen so quickly that it can't react in time and gets left behind. As a result, with their normal ego structure dissolved (at least partially), the person also experiences a sense of timelessness, or at least an acute slowing down of time.

This may also be why people who witness crimes usually perceive time as passing very slowly. Experiments have shown that when people see a fake assault, or when they watch a video of a bank robbery, they almost always massively overestimate how long the incident lasted, usually guessing between 250–500 per cent longer.[16] The sheer shock of witnessing these incidents may paralyse our ego too.

One query which might arise here is: don't we experience a state of egoless-ness in states of absorption and distraction too? Doesn't our thought-chatter fade away when we watch TV or when we're concentrating on tasks in the same way that it does when we meditate? And as a result, shouldn't we transcend our normal ego and experience a radically slowed sense of time, instead of a faster one?

But the experiences we're dealing with in this chapter are of a completely different order to states of absorption. The important fact is that in states of absorption the ego is not, strictly, in abeyance. It may not be very active at that moment, but as a structure it's still intact. There's still a sense of separation between itself and the world, even if at that moment the sense of separation may not be felt. But in these experiences of *intense* concentration we go beyond the ego completely; as a structure it fades away. The house of the ego is no longer standing. There is still a sense of identity, but this isn't the normal thinking ego. It's pure awareness without thought, the 'witness' (or *atman*) with which we are aware of the content of our own minds.

According to my Theory of Relativity, then, our personal sense of time is relative to two things: information processing, and the ego. Time goes more slowly a) as we absorb and process more information from our surroundings (or our thoughts); and b) as our ego becomes weaker.

And now that the second half of the theory is in place, we're in a good position to understand the unusual perceptions of time which occur in altered states of consciousness.

Time and drugs

Anybody who has ever taken psychedelic drugs will vouch for the time-expanding effect they have. Drugs like LSD and magic mushrooms can, in Aldous Huxley's phrase, 'telescope aeons of blissful experience into one hour'.[17] The Russian philosopher P.D. Ouspensky once made a series of 'mystical experiments', which probably (since he doesn't state which drugs he used) involved inhaling ether. He found that he could no longer understand the notion of 'separate existence', that 'all things were dependent on one another, all things lived in one another'. And he also experienced what he calls an 'extraordinary lengthening of time, in which seconds seem to be years or decades'. As he describes it:

> My companion was saying something. Between each sound of his voice, between each movement of his lips, long periods of time passed. When he had finished a short sentence, the meaning of which did not reach me at all, I felt I had lived through so much during that time that we should never be able to understand one another again, that I had gone too far from him ... There were no means of conveying to him all that I had lived through in between.[18]

A similar experience – this time with mescaline – is reported by the author W.T. Stace in his book *Mysticism and Philosophy*. Under the influence of the drug, the person – who we only know by the initials N.M. – experienced an extreme heightening of perception, in which 'every object in my field of vision took on a curious and intense kind of existence of its own ... Everything was *urgent* with life.' Along with this, N.M. felt that 'time and motion seemed to have disappeared, so that there was a sense of the timeless and eternal'. And later, as the drug was beginning to wear off, he or she had a strong sense of falling back into time. 'I began to be aware of time again, and the impression of entering into time was as marked as though I had stepped from air into water, from a rarer into a thicker element.'[19]

One reason why these drugs can bring about an expansion of time is because, in the phrase popularised by Aldous Huxley, they 'open the doors of perception'. By disturbing the normal chemical homeostasis of our bodies, they also, it seems, disturb the normal awareness-restricting mechanisms of our consciousness. The desensitising mechanism stops functioning, and as a result, we gain access to the intensely real world which we experienced as children. Or according to the psychologist R. Fischer, psychedelics bring an 'amplification of sensing, knowing and attending', and a 'torrential flood of inner sensations'.[20] This means that we process much more information than usual – or in Marianne Frankenhauser's terminology, there's an increase in the 'amount of mental content' – which expands our sense of time. In Ouspensky's case, the amount of perceptual and mental information he absorbed within a few seconds may have been as much as he would normally absorb in days, so that each few seconds seemed incredibly long.

However, the ego is an important factor here too. Psychedelic drugs are well known for the 'ego-subduing' effect they have. In fact in some cases the ego may actually dissolve away altogether, leaving a sense of nothingness which can be both frightening and liberating. The sense of being a separate I, complete with memories, opinions, personality traits, etc., fades away, and we may feel that we're fused with the world which normally seems outside us. And since our sense of time passing is connected to the development of our ego, it's clear that this adds to the time-expanding effect of psychedelics. By dissolving the ego, psychedelic drugs can take us 'out of time'.

The American psychologist Abram Hoffer experimented with LSD in the 1960s – before it was made illegal – and also experienced these effects. He described how a particular note of a song he was listening to became interminable, and left him with no recollection of its beginning or any anticipation of its end. On another occasion, he reported that 'while watching the pulsations of the electric clock, its second hand stopped moving and for a few seconds for me time stood still'.[21] More recently, the philosopher Benny

Shanon has exhaustively studied the effects of the ayahuasca drink used by the shamans and medicine men of the Upper Amazon, which contains the powerful psychoactive substance DMT. According to Shanon:

> The person under intoxication experiences lots of happenings and change, hence it seems to him or her that more time has passed than actually has. In extreme cases, it may seem that time has altogether stopped, or that temporal distinctions are no longer relevant.[22]

Interestingly though, other drugs don't appear to have the time-expanding effect of psychedelics. Another psychologist, Stephens Newell, made a long and systematic study of the effects of different drugs on time perception, and also concluded that 'the stronger psychedelics (LSD and others) have the effect of slowing or stopping time'. However, he also found that alcohol and heroin have the opposite effect, and actually make time pass faster.[23] Other studies have shown that, while LSD, marijuana and amphetamines usually expand time, tranquillisers and alcohol usually make people overestimate how much time has passed, as if it's moving faster.[24]

This might seem anomalous at first, but in fact it's exactly what we would expect, since these drugs are depressants. Rather than opening the doors of perception, they insulate us against reality, reduce our awareness, and therefore reduce the amount of information we process. It's perhaps significant that alcohol seems to have an ego-intensifying effect as well. As the sociologist John Archer writes in his book *Male Violence*, alcohol creates an 'accentuated feeling of power and self importance which makes ... one's own identity more easily threatened'.[25] It's logical to assume that, in the same way that time goes slower when our ego structure is weaker than normal, it passes faster when this is stronger than normal. And this may be the case with alcohol – our ego becomes more acute and we become even more subject to time.

Time and schizophrenia

The state of heightened awareness we experience when the 'doors of perception' are open isn't always a positive experience. New Age teachers often treat the ego as an enemy which has to be destroyed, but our real problem isn't the ego in itself, but that our egos have become too strongly developed and uncontrollable, making us too *separate* to the cosmos. It's true that it's our enemy in some ways, but we also *need* it. We need an integrated sense of self to make decisions, interact with other people, perform the practical tasks and chores of our lives, and so on. When we don't have a stable ego, or when the 'self-system' collapses altogether, this usually causes mental illness, with symptoms such as paranoia, delusions, a sense of unreality, and an inability to sense the boundaries of our bodies. As Ken Wilber has pointed out, destroying the ego doesn't lead to enlightenment, but to psychosis.[26]

However, this breakdown of the ego structure also often means that the desensitising mechanism stops functioning – so that, along with their other symptoms, sufferers from schizophrenia generally experience an intense state of consciousness. According to psychologists, one of the characteristics of schizophrenia is a 'heightened perceptual acuity', or a perception of 'sharpened intensity' with 'the raw data of new experiences'.[27] Or as Julian Jaynes puts it more dramatically: 'Schizophrenics are almost drowning in sensory data ... They seem to have a more immediate and absolute involvement with their physical environment, a greater *in-the-world-ness*.'[28] But although this may seem a positive thing, without a stable sense of self schizophrenia sufferers can't control or order their perceptions, or turn their attention away from them, and so they feel lost, stranded and even threatened by the intense reality of the world. As the psychiatrist J.S. Sullivan writes, 'the schizophrenic is surrounded by animistically enlivened objects which are engaged in ominous performances which it is terribly necessary – and impossible – to understand'.[29]

As a result of this, another of the normal symptoms of schizophrenia is a distorted sense of time. Sufferers often have trouble estimating the length of time periods, or distinguishing between events that have already happened and those they expect to happen. Their thoughts are disconnected and fragmented, without a logical sequence. And their lack of a normal sense of time means they struggle to organise their lives, often missing appointments and living without any kind of routine or structure.[30] But most significantly, time passes incredibly slowly to people with schizophrenia, because of the massive amount of perceptual information they absorb, and their weak (or completely absent) ego structure. As Jaynes writes: 'Patients may complain that time has stopped, or that everything seems to be slowed down or suspended.'[31]

In his essay 'Exploring Time in Mental Disorders', Dr A. Moneim El-Meligi gave two case studies of schizophrenia, paying special attention to the patients' perceptions of time. His first case, a 33-year-old 'hypomaniac' who worked over 80 hours per week, complained that , 'the day drags on so painfully that he avoids looking at clocks and watches'. El-Meligi asked him how fast time was passing that day and he replied: 'Today time has been funny, I guess it was weird. When you said it was noon, I thought it was only nine in the morning.'

His second case was much more serious – a 23-year-old American man who was severely depressed, and suffered from sensory distortions which meant that even simple actions like lifting a cup or drinking a glass of water were extremely difficult, since he'd see the cup or glass growing larger and larger. The man also told El-Meligi:

> Time is the worst thing. It seems so long that a month period is unimaginable. Time seems endless. *The languor of time* [El-Meligi notes] *makes it impossible for the patient to enjoy anything. He cannot, for example, follow a TV program or a baseball game.* You just cannot see the end of it. It seems so long. I hate time, it seems very real. I make myself sleep in order to conquer time, but it does not work.[32]

Another factor here may be that people with schizophrenia process a much larger than normal amount of mental information. Their minds seem to run faster, with even more chaotic thought-processes than 'normal' people's brains. The following description of a psychotic episode shows this clearly.

> Inner experiences took place at such an increased speed … much more than usual happened per minute of external time. The result was to give an effect of slow motion … The speeding up of my inner experiences provided in this way an apparent slowing down of the external world.[33]

Hypnosis and sleep

On one level hypnosis is a perfect example of the working of the third law of psychological time ('Time passes quickly in states of absorption'). As in the trance-like state we sometimes get into when we watch TV, under hypnosis the amount of perceptual and thought information which we process drops to an extremely low level. We experience a state of intense absorption, with our attention narrowed to the instructions of the hypnotist and the tasks he gives us to do. It's therefore not surprising that as Roy Udolf – author of *Handbook of Hypnosis for Professionals* – reports, many hypnotic subjects underestimate time, usually by around 40 per cent.[34] In fact, some therapists recommend hypnosis as a way of making unpleasant situations pass quicker. The hypnotherapist Paul Gustafson, for example, recommends it for visits to the dentist. As he writes: 'Hypnosis can help dentistry clients as an effective analgesic adjunct. It relieves anticipatory anxiety [and] distorts time perception, speeding up the procedure.'[35]

Perhaps this is only what happens in a more 'shallow' state of hypnosis though, when the structure of the ego is still intact. There appears to be a deeper level at which the ego dissolves completely away, and in which time slows down drastically, or completely disappears. At this level, it only takes a simple suggestion from the hypnotist to take you into a timeless realm.

In an experiment carried out by the psychologists Barber and Calverley in 1964, for instance, a hypnotised group of people were given some 'time-slowing' suggestions, and then asked to learn a list of twelve nonsense sounds. At the end they were asked to estimate how long they'd spent doing it. The actual time was 5 minutes, and while a control group of another sixteen non-hypnotised people (who weren't given the syllables to learn) estimated an average of 4.2 minutes, the hypnotised people estimated an average of 89.1 minutes.[36]

The most striking pieces of evidence for the time-slowing effect of hypnosis, however, are the amazing feats which hypnotised people are capable of, which (like those performed by sportspeople in the Zone) would be impossible in normal time. The psychologist Gordana Vitaliano gave hypnotised subjects a complex mathematical task, which normally takes around 10 minutes, but which they completed in just 15 seconds.[37] In another experiment, a hypnotist slowed down his subjects' sense of time by pretending to slow down a metronome until they believed it was only beating once every minute. He then told them they had an hour to accomplish certain mental tasks, such as planning a complicated meal. In reality they were given just 10 seconds, but they fed back the kind of wealth and detail of information which would only have been possible after an hour or so of normal time.[38] Similarly, a hypnotised woman produced several dress designs in less than a minute, saying afterwards that she felt like she had experienced an hour or more of normal time within that minute.[39]

In fact these effects are so easy to create that some hypnotherapists and NLP (Neuro-Linguistic Programming) teachers use them as 'accelerated learning techniques'. By using self-hypnosis, they tell us, we can create a state of 'time distortion', in which we can review or absorb massive amounts of information in seconds.

Sleep is a egoless state too, of course – in fact the ego (or normal conscious mind) is the very part of us that sleeps. In the last chapter I suggested that in sleep we normally have the sense that time passes quickly, because of the lack of information our minds process. Non-dreaming sleep is

similar to a coma – a state of absence, or suspended animation. But when we dream – usually for around 1.5 hours a
night – this changes. Often the passing of time in dreams
approximates fairly well to our normal experience, but
sometimes it expands massively in a similar way to deep
states of hypnosis. (Incidentally, you might question
whether dreaming can be an egoless state. Isn't it the ego
which is doing the dreaming? But as I mentioned earlier in
this chapter, there's an awareness inside us which exists
apart from the ego. We can still process information and be
aware of sensations even if the ego is absent.

A few years ago, I fell asleep during a long coach journey
in Malaysia, and had a dream about being back at my university. It wasn't particularly interesting (my wife always
tells me that my dreams are incredibly dreary) – for some
reason, I'd decided that I wanted to play for the university
football team, and turned up at a training session, only to
realise that I'd forgotten my football boots. And then I
realised that the other people in the team were people I did-
n't like from my hall of residence. I played the match
anyway, but the people from my hall kept trying to tread on
my toes and my feet kept sinking in the mud. And then I
had to go into the showers with them after the match ... I
won't go on, but the main thing is that to me the dream
seemed long and intricate; it covered perhaps three hours of
clock time, and I felt as though I'd experienced all of that
time. Then I was woken up by the sound of loud crackling
as the coach driver turned on the television, which wasn't
tuned to a station. I was puzzled by the fact that we still
seemed to be in the same town as when I'd fallen asleep,
and I asked my wife how long I'd been asleep for. I was startled when she said: 'Not long – about 5 minutes.' Similarly,
the 19th-century French philosopher Jean-Marie Guyau told

the story of a student who fell asleep for a few seconds, before his friends woke him up. In those few seconds he had a dream about going to Italy, touring around different towns and visiting different monuments and meeting a lot of people. He experienced so much that he felt as though he'd been asleep for hours.[40] I also once heard a composer say that he was inspired to write one of his pieces by hearing a peal of thunder in the middle of the night, when he was on the verge of sleep. The thunder could only have lasted for a couple of seconds, but to him it seemed to go on forever, to stretch out for at least several minutes, and he aimed to reproduce its long, roaring monotone in his music.

These effects are very similar to the massive expansion of time that can take place under hypnosis, and are clearly good examples of the workings of the fifth law of psychological time. In Chapter 7, we'll see that both hypnosis and sleep can apparently affect our sense of time in another striking way: namely, by enabling us to slip out of linear time and experience the future and the past in the present.

Other conditions

So far in this chapter we've been looking at states in which the ego is in abeyance, when it's paralysed by shock, transcended by intense attention, temporarily dissolved by drugs or hypnosis, or permanently dissolved by mental illness. But it's also a question of degree, of course. The ego doesn't have to be completely paralysed or dissolved; it may only be *partially* disrupted, causing a less extreme disruption to time perception.

This seems to the case with autism, for instance. People with autism often have problems dealing with time. Psychologists describe them as being 'lost in time', and lacking the intuitive sense of time which makes most of us able to roughly guess the time and estimate how long events will last.[41] They have difficulty understanding the concept of time periods – minutes, hours and days, for example – and don't have a 'feel' for how long an event will last, or how long they have to wait until something happens. A person who suffers from Asperger's syndrome (or 'high-functioning' autism, as it's sometimes called) told me: 'I have no concept of time at all. I have absolutely no idea how long things last or how long it is before I have to do something else. Without a watch I'm completely lost. And even if I wear one I'm still usually late for everything.'

And interestingly, autistic people seem to be more affected by the fourth law of psychological time ('Time passes quickly in states of absorption') than other people, perhaps because it's easier for them to reach a very intense level of absorption. When they focus their attention completely, they may not just find that time goes quickly, but that it disappears altogether. Here an autistic person tells the story of how he experienced time when he was working at what he calls 'a very menial data-entry type job'.

> It used to be really scary … I would check the hard copy and ensure it was entered in the system correctly. EVERY day, I would at some point take a breath, blink, and then realize 6 hours had passed without me realizing it, and that I could not remember a single event that had happened during that time. I knew I had caught errors and corrected them, and the pile of hundreds of files had been gone through, but I could not remember anything about the last several hours. It was even scarier when I found that I NEVER had errors when this happened. It became a completely robotic experience.

In a similar – but less severe – way, people with dyslexia often struggle with time. When people are assessed to try to find out if they're dyslexic, the assessor usually asks them about their sense of time – if they find that they're often late for appointments, if it's difficult to organise their lives in terms of time, if they had difficulties learning to tell the time as children, and so on. Along with clumsiness, a poor sense of direction, difficulty telling left from right, a poor sense of time is one of the typical signs of dyslexia. It seems that, as with autism, people with dyslexia don't have an intuitive sense of how long periods of time are. As the author of *The Gift of Dyslexia*, R.D. Davis, points out, their sense of time is distorted, so that they never experience it in a uniform way. As he writes: 'A minute can be a very long time or very short – but it is never the same.'[42]

Unfortunately we still don't know exactly what it is that makes autistic and dyslexic people different. The psychologist Simon Baron-Cohen describes autism as an 'empathy disorder', which makes a person unable to put himself (the vast majority of autistic people are male) 'into other people's shoes'.[43] Autistic people are self-immersed and detached; they find it difficult to communicate and don't understand other people's emotions or reactions. Baron-Cohen suggests that most men have these characteristics to a lesser degree (after all, these are women's most common complaints about us), and that the autistic mind can be seen as an 'extreme male brain'. He believes that autism is a condition that occurs when the typically 'male psyche' develops to an extreme level.

One theory about dyslexia is that it's related to the two hemispheres of the brain and their different functions. The hemispheres seem to work in different ways for dyslexic people. A study by the National Institute of Child Health and Human Development in Washington showed that dyslexic children had a less well-developed left hemi-sphere. In particular, the rear portion of this hemisphere – which is associated with logical and sequential thinking, and processes language and communication – was signifi-cantly smaller than normal.[44] In other words, the 'left brains' of dyslexic people don't appear to function as well as they do for most of us. We saw earlier that our sense of time appears to be a left-brain function, so it makes sense that dyslexic people should have problems with time. However, in compensation for this, their more dominant right brains mean that they have better than usual spatial awareness, visualisation and intuitive thinking – which is why so many dyslexic people are creatively and artistically gifted.

The important point here is that, with both autism and dyslexia, the self-system doesn't develop in the normal way and is disrupted (if to different degrees). And since time perception is so closely linked to the self-system, it's inevitable that this also becomes distorted slightly.

Positive wakefulness

The same is true of all of the experiences we've looked at in this chapter: they all show how bound up our sense of time is with the ego. As soon as the ego is disturbed in some way, or put into abeyance, then time slows down radically or dis-appears altogether. This suggests that in a sense our experience of an evenly flowing linear time is a kind of construct – even a kind of illusion – created by the ego. For children, time doesn't exist *before* the ego develops, and the evidence we've looked at in this chapter suggests that it doesn't exist *outside* the ego, in the same way that the seeming reality of a dream ceases to exist the moment you wake up.

So here we have the beginnings of an answer to the ques-tion which is at the heart of this book: how can we make ourselves less subject to time, and stop it passing so quickly in our lives? If we know that time passes so fast because we have strongly developed egos and because we absorb very little perceptual information as we go through our lives, then we need to find a way of subduing or weakening our normal ego and of 'waking up' to reality so that we absorb more perceptual information.

But hold on, you might be thinking, are you suggesting that we take psychedelic drugs, or stage our own mental breakdowns so that we become schizophrenic – or even put ourselves in a permanent state of hypnosis? It's true that if you did this you'd probably live through the equivalent of thousands of years of normal time – perhaps even more. But the drawbacks would be so great that it wouldn't be worthwhile, of course. In fact 'you' wouldn't even be there to be aware of the aeons of time you were living through, since both hypnosis and schizophrenia involve a dissolution of the normal self-system (although this is only temporary in the case of hypnosis), as does the long-term regular use of psychedelic drugs. In any case, nobody would ever think it was worth enduring the torments of schizophrenia just to live through more time, just as nobody would want to endure a life of constant boredom, even though that would make their lives last much longer.

But this doesn't mean that it's impossible to live in a state of the kind of wakefulness or egoless-ness which slows down time. Drugs and schizophrenia are a kind of negative heightened awareness, which only comes at the expense of psychological health. But there's also a positive equivalent to these, a way in which it's possible to live in a state of heightened awareness and with a weaker ego structure (and therefore an expanded sense of time) without these bad side effects – as we'll see in the next chapter.

5

Time Across Cultures

Time isn't just relative to different states of mind; it's relative to different cultures too.

Westerners tend to see time as a linear process, as a kind of river which flows from the past through the present to the future. The past is a bottomless trash box of moments which were once present but have now disappeared and will never be known again, while the future is an endless sequence of moments waiting to become present, which are unknowable until they arrive. This idea of linear time runs through Judaic and Christian beliefs, with the notion that the world has a beginning and an end, and that time itself was created with the world and will end when the world ends. As St Augustine wrote: 'The world was made with time and not in time.'[1] The Christian Church's hostility to cyclical views of time was a part of their general hostility to paganism. Cyclical time is rooted in nature, in the cycles of day and night, the moon, the seasons and the year. Christianity replaced nature-based religions with the worship of a transcendent God who is outside nature, and rejecting the cyclical view of time was part of this process. The Church also had to establish a 'transcendent' view of time which was not tied to nature. St Augustine stated this clearly. 'It is only through the sound doctrine of a rectilinear course [of time] that we can escape from I know not what false cycles discovered by false and deceitful sages.'[2]

Even now many scientists put forward a strictly linear view of the universe, telling us that it was created in a Big Bang about 16 billion years ago, when a giant mass of matter exploded out and slowly formed into galaxies and stars and planets. The force of the Big Bang means that all galaxies and stars are still expanding out into space. But at some point in the future, when that initial exploding force has died down, gravity will begin to pull all of them towards each other again, and eventually the whole universe will collapse in on itself in a 'Big Crunch' and cease to exist. (Although other scientists believe that the universe will simply keep drifting apart forever.) In other words, time has a beginning and an end, just as in the Judaic/Christian worldview.

In historical terms, though, this view of time is an anomaly. By far the majority of the world's cultures have not had a linear sense of time. The Maya of South America believed that patterns of events repeat themselves every 260 years, as a part of 'deep time cycle' which lasts around 5,000 years. The present cycle is – rather scarily – due to end in 2012, but this doesn't necessarily mean the world will end then. To the Maya, the concept of the 'end of the world' has no meaning, since the world has been created and destroyed many times, and this cycle will continue forever. The ancient Greeks conceived of a 'great year' of 36,000 solar years. Some Greek schools believed that history would repeat itself in exactly the same way, with exactly the same events and people – the Stoics, for example, called this endless repetition *palingenesia*. But more commonly, the idea was that each cycle would move through the same phases, with the same *pattern* of events. Similarly, the ancient Hindus believed that history consists of four ages (or *mahayugas*), during which certain characteristics and certain kinds of behaviour are prevalent. These last for a total of almost 4.5 million years and then repeat themselves. However, even these are only a small sub-cycle of the overall cycle of the universe, which is called *Kalpa*. This covers the birth of the universe to the point where it returns to the unmanifest state which it emerged from. Each *Kalpa* lasts for a total of 1,000 *mahayugas*, which is almost 4.5 billion years. At this point there is a pause for another 4.5 billion years, then the universe becomes manifest again and the cycle of history starts anew.[3]

Timeless cultures

The idea of cyclical time still implies a flow though. It's just that rather than flowing on indefinitely, time flows to a certain point and then switches back to the beginning. However, there were some indigenous peoples to whom, it seems, time hardly seemed to flow at all.

I'm using the past tense here because unfortunately all but a small minority of the world's indigenous cultures have now been disrupted to some degree through contact with Western cultures. In most areas of study, research becomes less important the older it is, but this doesn't strictly apply to anthropology. In fact, although older anthropological studies may be coloured by the prejudices of different decades, they can often be more valid than contemporary studies, since they often deal with cultures closer to their original pre-contact state. This process of cultural disruption has continued at devastating speed over the last 30 to 40 years or so, and has been one of the most shameful aspects of the spread of Western capitalism around the globe. (This is why many of the anthropological studies I refer to in this chapter are from the early to mid-20th century.)

Many anthropologists have been struck by the incredible patience which indigenous peoples appeared to show. When the anthropologists Andrew Miracle and Juan de Dios studied the Aymara Indians of Chile in the 1970s, for example, they were amazed that Aymara people were seemingly content to wait for a whole afternoon for a truck to come and take them to market. They never showed the slightest trace of frustration, no matter how long they had to wait.[4] Similarly, when the anthropologist Edward T. Hall (author of *The Dance of Life*, which I've already quoted from) began to work on Native American reservations in the 1930s, he noticed that Indians never seemed to be disturbed by waiting. Whereas it's normal for Europeans to fidget impatiently and become irritable, the Native Americans he saw waiting at trading posts and hospitals never showed any sign of irritation whatsoever, even if they had to wait for hours. As he writes:

> An Indian might come into the agency in the morning and still be sitting patiently outside the superintendent's office in the afternoon. Nothing in his bearing or demeanour would change in the intervening hours ... We whites squirmed, got up, sat down, went outside and looked toward the fields where our friends were working, yawned and stretched our legs ... The Indians simply sat there, occasionally passing a word to one another.[5]

Of course, these are only the observations of European-descended anthropologists – how could they be sure what the Native Americans were thinking and feeling? Perhaps the Indians were irritated by waiting but were just good at disguising it. However, another possibility is that they weren't irritated because they didn't have the Western sense of linear time, with its concept of 'precious' moments passing by and being wasted. Westerners hate waiting so much partly because we see it as time 'wasted', and partly because we're acutely aware of a future event which we want to reach as soon as possible. We resent the moments that are separating us from this future event and want to get them out of the way as quickly as possible. But perhaps, if you don't isolate time into past, present and future moments you'd have a different attitude to waiting. Perhaps you'd simply rest in the 'now', without feeling such a pull towards the future.

This is the conclusion Hall himself came to. After living with the Hopi and the Navajo in Arizona from 1933 to 1937, he came to believe that the two tribes live in a kind of eternal present. He noted that the Hopi language doesn't contain a word for time, and their verbs have no past, present or future tenses. 'To the Hopi', he wrote, 'the experience of time must be more natural – like breathing, a rhythmic part of life'.[6] Hall learned that the language of the Sioux has no word for time either, nor words meaning 'late' or 'waiting'. The Sioux 'don't know what it is to wait or to be late', he was told by a Western-educated Sioux.[7]

Another American anthropologist, Barbara Tedlock, lived with the Quiche Indians of Guatemala (who are descended from the Maya) for several years in the 1970s and also concluded that they don't have a linear sense of time. Whereas we isolate time into 'moments' that come and go, she noted that for the Quiche, 'at no given time, past, present, or future, is it possible to isolate that time from the events that led up to it'.[8]

This seems to be true of many of the world's indigenous peoples. Traditionally, the concept of time had little or no meaning to them; very few of them appear to have even had words for time, or for the future or the past. As the anthropologist Robert Lawlor noted in his book *Voices of the First Day*, none of the hundreds of different Australian Aboriginal languages had a word for time, nor do any traditional Aboriginal peoples have a concept of time.[9] In the 1930s, the English anthropologist E.E. Evans-Pritchard lived with the Nuer people of southern Sudan, resulting in one of the most famous of all anthropological texts, *The Nuer: A Description of the Modes of Livelihood and Political Institutions of a Nilotic People*. Evans-Pritchard also described the Nuer as living 'out of time'.

> The time perspective of the Nuer is limited to a very short span – in a sense they are a timeless people … They have no word for time in the European sense. They have no conception of time as an abstract thing which can be wasted, or saved, or which passes.[10]

Counting out time

In contrast, the Western sense of linear time is so strong that it's important for us to name and number the different intervals of time that we live through. We always know where we are in time – how old we are, and what day of the week and what year it is. It's as if our lives are a journey and we need to be constantly aware of how far we've gone, and where we are along the way. But many traditional indigenous cultures don't appear to have had this need, and didn't number years as they passed, or know their own ages in terms of years.

Another influential early anthropologist, Irving Hallowell, lived with the Saulteaux Indians of Northern Canada during the 1930s, and found that they barely used the time unit of a year at all, and didn't think of one year as containing a specific number of days. Rather than thinking of a person as being a certain number of years old, they had much broader categories of age, such as young child or adolescent.[11] The anthropologist Emiko Ohnuki-Tierney studied the Ainu people of Northern Japan in the 1960s, in a remote area where their culture had yet to be severely disrupted, and found that they didn't distinguish different days of the week. The only kinds of time markers they used for days were words for tomorrow and the day after tomorrow, and yesterday and the day before. They didn't count the years or track their ages either, and the only kind of markers they used for past events were other events that happened at the same time, or natural phenomena that recur each year. In other words, rather than saying 'my brother died in 1992', an Ainu – or a Saulteaux – might say something like 'my brother died at the time of forest fire', or 'at the time when we cleared the plot'. And instead of saying 'my child was born in February', they would say something like 'my child was born at the time of the first frost', or 'at the time of the herring run'.[12]

This highlights another major difference between our concept of time and native peoples'. For us time is an abstraction. The seconds, minutes, hours and weeks which we use to measure our lives don't correspond to any natural phenomena – they're just man-made mathematical divisions of time. Days, months and years are real phenomena, of course, but even they are abstract concepts in the sense that we've given them different names, and think of them in terms of sequence. In reality there's no such thing as 'Monday', 'February' or a year called '2006' – the earth just keeps spinning on its axes, the moon rotates around the earth and the earth rotates around the sun.

But for many indigenous cultures, time could never be separated from nature. The only kind of dating system used by the Trobriander islanders of Oceania, for example, was tied to their yearly cycle of gardening. They'd say that a particular event happened – or was going to happen – during the clearing of the shrub, in burning, in planting time, during weeding, the trimming of the vine or during the harvest proper. However, they didn't have a generally agreed point when one year ended and another began. There was no need for this, since – like Ainu and the Saulteaux – they didn't think in terms of a sequence of years.[13]

Similarly, whereas European months are mostly named after obsolete Roman gods, the Khant people of northern Siberia named theirs after the activities and natural phenomena that occur during them. They have months called the Naked-Tree month, the Pedestrian month (when people can't travel on horseback but only on foot because of ice), the Crow month, the Pine-Sapwood month, and the Salmon-Weir month.[14]

In other words, for these peoples time didn't exist as an independent concept, but only as an aspect of the natural world, in terms of the yearly cycle of natural events or the duration of natural processes. In Western cultures, however, time is divorced from nature, in a way which mirrors our general sense of alienation from the natural world.

One consequence of this is that, whereas we let our artificial time tell us when to begin activities, indigenous peoples would typically wait until the time was *right* before starting something, either until natural phenomena give them a particular signal, or until they intuitively *felt* that all the conditions were right. For example, the Leco-Aguachile people of Bolivia would never set a particular time to start fishing, but waited until bats hover over the water, while the San people of the Kalahari Desert in southern Africa would only begin hunting once they'd assessed the environment and the behaviour of animals and were sure that it was the 'moment to be lucky'. As the contemporary author Jay Griffiths – from whom I have taken the last two examples – puts it: 'For every indigenous group I have ever known or read about, timing in social interactions is indeterminate, unpredictable, demanding flexibility.'[15]

In view of this, it's perhaps not surprising that many early anthropologists found that indigenous peoples had a very lax attitude to what we would call 'punctuality'. Punctuality means being dominated by artificial time, letting time tell you what to do rather than circumstances. And to many indigenous groups the Western concept of being 'on time' had no meaning whatsoever. Irving Hallowell described how the Saulteaux's ceremonies never had a fixed time; the leader would simply start drumming when he felt ready, and people would arrive at any time over the next few hours. Hallowell also noted that if he tried to fix a time to meet with one of his Native American informants, they usually kept him waiting for hours.[16] Similarly, Edward T. Hall noted that to the Pueblo people 'events begin when the time is ripe and no sooner'. He told the story of going to a Christmas dance near the Rio Grande, and waiting for hours for it to begin. According to Hall, the other white visitors muttered impatiently to themselves, shivering and stamping their feet, while the the Pueblo people there showed no sign of urgency whatsoever.

And eventually, in the middle of the night, when the whites were on the verge of exhaustion, there was an eruption of drums and rattles and deep male singing voices.[17] Again, you might say that this was just Hall's assumption – that he had no way of knowing what the Indians were actually feeling inside, and that perhaps they were impatient but keeping it hidden. But it certainly suggests that this culture was free of time pressure, and that – like the San or Leco-Aguachile peoples – they waited to begin events until they intuitively felt that conditions were right.

The sense of the past and the future

The reports of anthropologists suggest that indigenous peoples were much less concerned with the future and the past than Europeans. As we've already seen, our sense of the future and the past alienates us from the present – we spend so much time thinking about them that we forget about the present that we're *actually* living in. But traditionally, many indigenous cultures were much more 'present-centred'. Some cultures were so unconcerned with the future and the past that they scarcely distinguished between them. In the dialect of the Inuit of Canada's Baffin Island, for instance, the same word – *uvaitiarru* – is used for the distant past and the distant future.[18] When Edward T. Hall worked on Hopi and Navajo reservations in the 1930s, he found that the concept of future deadlines meant little to the Indians, so that houses that should've been finished years ago were still without windows or roofs. Dams that were supposed to be finished within three months were still unfinished after a year. He noted that if you told a Navajo he could have your best race-winning horse in the fall he'd be disappointed and walk away; but if you told him that at this very moment he could have your worn out old horse with knock knees and pigeon toes he'd be delighted.[19] Perhaps Hall was being a little naïve, and the real reason why these Native Americans were less than enthusiastic about completing these projects was because they were a part of a culture which was oppressing them, or simply because they were culturally foreign to them. However, taken together with his other observations of their attitude towards time, Hall was convinced that this showed a less future-oriented mentality. As he wrote: 'To the Navajo the future was uncertain as well as unreal, and they were neither interested in nor motivated by "future" rewards.'[20]

Part of the problem, Hall believed, was that the European sense of linear time means that we are, in his term, 'monochronic'. This means that we usually do one thing at a time, that when we have a task to do we concentrate all our attention on it and try to complete it in the shortest possible time, before moving on to the next thing on our list of things to do. But according to Hall, the Hopi and Navajo operated by 'polychronic' time, which meant that completing one particular task wasn't so important to them, and that they were happy to do several different things at once.

The reports of anthropologists suggest that, in general, the concept of the past had little meaning to indigenous peoples either. As the anthropologist Maurice Bloch noted, African peoples such as the Hazda and Mbuti never talk about the past, and have no concept of 'history'.[21] Many indigenous peoples don't have a tradition of oral history which is passed down from one generation to the next, and seem to have very little concern for past events or important personalities from the past. Although most of them have origin myths of one form or another, these aren't history in our sense of the term. They aren't narratives like Western (or Middle Eastern) creation myths, describing a process of one thing being created after another, and the creation of the world isn't necessarily seen as complete or as a past event. As another prominent American anthropologist, Elman R. Service, wrote of the Trobriand Islanders (unfortunately using the pejorative term 'primitive'):

> The Trobriander is distinctly unhistorical in his cosmogony. Agriculture, magic, law, the people themselves, islands – everything – came into being once, full-blown, and that is the way they *are*. The Trobriander, like so many … [indigenous] peoples, does not view phenomena as being in a process of change through time. A being is changeless; therefore, the Trobriand language has not word for *to be* or *to become*.[22]

The Pirahã people of Brazil don't have any creation myths, or any sense of the past at all. When they're asked about how the world came into being, or what happened at a certain point in the past, they just say: 'Everything is the same, things always are.' According to the ethnologist Daniel Everett, who has lived with them for a total of seven years (from the late 1970s to the present) and made a detailed study of their language, the Pirahã have no past tense verb forms, and hardly any words associated with time. They don't appear to have a sense of sequence at all; they don't use clauses to indicate chronology (e.g. phrases like 'when I have finished hunting, I will go home') and never use numbers. Everett believes that they have no interest in the past because they live so fully in the present; if something isn't happening right at this moment then it has no value and there's no point talking about it. As he puts it: 'All [their] experience is anchored in the present.'[23]

According to Irving Hallowell, the Saulteaux did have some sense of history, but this only stretched to about 150 years ago. Anything beyond that belonged to a mythological past in which events weren't defined in terms of time. The Saulteaux spoke of a time long ago when the earth was young and giant animals lived together with mythical beings, but these events didn't have any set sequence. As Hallowell noted, for them, 'all means ... of temporal orientation are *local*, limited in their application to the immediate future, the recent past, immediate activities'.[24]

In the traditional cultures of Australian Aboriginal peoples, there's also no clear distinction between the present and the past. According to their mythologies, the world was created during the Dreamtime, when giant beings strode over the earth's surface, leaving their imprints as mountains, lakes and oceans and the rest of the earth's topography. But the Dreamtime isn't a historical period which happened long ago and is over and done with. According to the Australian scholar Paul Wildman, in a sense the Dreamtime is still happening now, and the world is still in the process of being created.[25]

Individuality and time

So why does time seem to be less meaningful as a concept to indigenous peoples? Why don't they appear to have a sense of linear time or arbitrary naming systems for units of time? And why do they seem to be so unconcerned with the future and the past?

This may well be related to the fifth law of psychological time. We've seen that in Western society our strongly linear sense of time is the product of our ego, that it develops in parallel with the ego as we grow into adulthood. And perhaps the reason why indigenous peoples have a less defined sense of time is because they have a less sharpened and separate ego than us.

In my book *The Fall*, I try to show that this is the essential difference between modern Eurasian peoples and indigenous peoples in traditional cultures. Indigenous peoples didn't exist as individuals in the sense that we do; their identity couldn't be separated from other people or nature. The community and the land were parts of their *being*, extensions of the self. The early French anthropologist Lucien Levy-Bruhl cited reports of indigenous peoples who use the word 'I' when speaking of their group,[26] while the contemporary anthropologist George B. Silberbauer notes that, to the G/wi people of the Kalahari, 'identity was more group-referenced than individual. That is, a person would identify herself or himself with reference to kin or some other group.'[27] Writing in 2002, Dr Spike Boydell observed that even now the indigenous peoples of Fiji have a similar concept of the 'self-embedded-in-community [which] contrasts with the western value of individualism with its idea of the self as separate and separating from others'.[28]

In the same way, for indigenous peoples their land is a part of their very identity, as much as part of their being as their own body. This is one of the many reasons why the forcible relocation of indigenous peoples by governments is such a tragedy. Their attachment to their traditional land is so powerful that they experience this as a kind of death. The contemporary Fijian-born anthropologist A. Ravuva, for example, notes that the Fijians' relationship to their *vanua* or land is 'an extension of the concept of self. To most Fijians the idea of parting with one's *vanua* or land is tantamount to parting with one's life.'[29] It's therefore not so surprising that some contemporary indigenous peoples – such as the U'wa people of Colombia or the Kaiowa of Brazil – have threatened to commit collective suicide if their land is taken from them. As one Kaiowa chief said: 'We *are* the land.'[30]

The naming practices of some peoples suggest this sense of collective identity too. Indigenous peoples' names are rarely fixed and permanent. The names of Aboriginal Australians, for example, could change many times throughout their lives, often including the names of other members of their tribe. Other indigenous peoples use tekonyms – terms which describe the relationship between people – which change as relationships change. For example, in Bali women are often called 'mother-of ___', men are called 'father-of ____'. But when a grandchild is born their names change to 'grandmother/grandfather-of ____'.[31]

Writing in 1975, the American anthropologist Alvin M. Josephy described how this idea of collective culture was still a strong part of Native American life: 'Many Indians still do not understand or cannot accept the concept of private ownership of land ... Many find it difficult, if not impossible, to substitute individual competitiveness for group feeling.'[32] In contrast, individuality is the very basis of the Western way of life. A person has their own name, house, job, possessions, and so on, and is meant to compete against other individuals for jobs, money and status. This is undoubtedly one of the reasons why many indigenous peoples have had such difficulty adapting to the Western way of life.

When English missionaries in Australia sought to Europeanise the indigenous peoples, they would try, as the anthropologist Bain Atwood describes it, to make each individual 'an integrated centre of consciousness, distinct from the natural world and from other aborigines'. In order to do this, they would forcibly separate families and communities from each other and from their land. They also used the baptism ritual to enforce permanent English names, instead of the fluid Aboriginal names, to try to increase a separate sense of self. However, the Aboriginal people's sense of collective identity was so strong that these attempts almost always failed: they constantly swapped possessions and went in and out of each other's houses as if they had no concept of individual ownership at all.[33]

And so, in the same way that the Western sense of linear time, along with an acute awareness of the past and the future, is the result of a strong ego structure, indigenous peoples' less defined sense of time may be the result of their weaker ego structure. However, this doesn't mean that indigenous peoples were living in a *completely* timeless world. To a certain degree they were still subject to time – living their lives in a linear fashion, with some sense of sequence and of the duration of events. It's just that their sense of time was far less rigid and defined, and the future and the past were neither as real nor as important to them.

Thought and time

This brings us back to the relationship between the ego and linear time. So far we haven't looked at the question of *how* our strong ego produces a linear sense of time.

According to the American philosopher Ken Wilber, this is because of the intensified fear of death that the separate sense of self brings. If you live in a state of increased separateness, your existence is now more valuable and important to you, and you fear its end more. And because of this fear we have to *deny* death, to pretend that it's not going to happen. We have to convince ourselves that, rather than there being an end to time, we have an endless future in front of ourselves – in other words, that there is an afterlife, in which we can live forever. And this only becomes possible if time is linear. And in addition, Wilber says, the notion of an everlasting future gives us the opportunity for endless fulfilment of the desires for power and wealth, which the development of our strong ego structure also gave rise to. As he writes:

> In its more intensified awareness of death, the ego needed more time ... By cutting itself loose into a linear and progressive world of time ... the ego's essentially unquenchable and unfulfilable desires had room to pitch forward everlastingly.[34]

However, I believe that the relationship between the ego and time is more direct than this. In my view it's a simple question of *thought*. Our strong ego structure gives rise to what I referred to in Chapter 3 as our 'thought-chatter', the habit of constantly talking to ourselves inside our heads. In order to think, there has to be an 'I' to think with; and the more individualised that 'I' is, the stronger this 'thought-chatter' is likely to be. We seem to have lost control of this mental talk; our egos chatter away automatically, leaving us with one of our biggest problems: the fact that our egos won't *stop* talking to themselves, keeping us awake at night and creating a sense of mental chaos and triggering negative emotions. (As we saw in Chapter 3, this is the reason why time passes so slowly in states of non-absorption.)

And in my view, it's this constant stream of thought-chatter that gives rise to our strong sense of linear time. Almost all of it is concerned with the future and the past in some way – memories of past experiences, or plans, daydreams and projections of the future. This is inevitable because thinking can only deal with abstractions or ideas, and the past and the future *are* abstractions. It's almost impossible to think about the present, because it's there in front of you, as a reality. As St Augustine said, 'The past is only memory and the future is only anticipation', and we *create* time through the memories of the past and the anticipations of the future which constantly flow through our minds.

On the other hand, indigenous peoples probably weren't as prone to this thought-chatter. With their weaker egos, they wouldn't have had a constant stream of thoughts about the past and the future in their minds (at least not to the same extent). As a result, to them the only 'real' tense was the present, and there was little sense of time flowing, or of the future and the past. You could compare this to how we experience time when we manage to stop – or at least slow down – our thought-chatter during or after meditation. If you have a good meditation, there's almost always a powerful sense of nowness. You find yourself spontaneously living in the moment, giving full attention to your surroundings and to the things you're doing. You find that you seem to be free of the niggling worries and anxieties which normally occupy your mind – worries about the day ahead or about embarrassing things that happened yesterday. And this is precisely because your normal thought-chatter has slowed down. The future and the past no longer have any significance, because your thought-chatter is no longer creating them. You're no longer caught up in linear time. And this may be analogous to the way in which indigenous peoples are less 'time-bound' and more oriented in the present.

And of course, as we saw in Chapter 1, this is how our linear sense of time develops as we grow into adulthood. Wilber might say that our individual 'fall' into time as we grow up is connected to awareness of death too. It may be that as we develop an 'intensified awareness of death', we project an endless future of linear time in front of us as a compensation. But in my view it's more likely that this is due to the habit of 'self-talking' which develops as we become adults. Children have a 'fuzzy' sense of time because of their undeveloped egos; they have very little sense of the future or the past because their minds are largely free of thought-chatter. But by the time we reach the age of sixteen or so, we have become fully self-conscious and self-reflective, and as a consequence we have begun to – as Pascal put it – 'wander about in times that do not belong to us'.

Does time move slower to indigenous peoples?

Another interesting point about indigenous peoples' sense of time is that, if they have a weaker ego structure than us, they must also presumably have a *more expansive* sense of time. I've suggested that it's partly because children have a less developed ego that time moves so slowly for them, and that time slows down or disappears in states of ego-abeyance caused by accidents or intense concentration (such as 'Zone' experiences). And so, if the fifth law of psychological time is true, it must apply to indigenous peoples too. Time must be moving slower to them.

This might sound a little bizarre, but there's another aspect of indigenous peoples' psyche which points to this conclusion too: their 'heightened' perception of their surroundings. In Chapter 2 I suggested that the reason why time seems to speed up for us is that as we get older our perceptions of our surroundings and our experience become more automatic, more 'familiarised', so that we process progressively less information. But it may be that, for traditional indigenous peoples, this process of familiari-sation didn't occur, or at least not to the same degree.

As I mentioned in Chapter 2, our ancestors in central Asia and the Middle East were subject to very severe survival pressures, particularly from around 4000 BC. At that time most of the world's population were still living as hunter-gatherers, and life wasn't particularly difficult for them. So perhaps, while our ancestors developed a 'de-sensitising mechanism' as a survival aid, other peoples – who later became the indigenous peoples we're familiar with – didn't need to.

In 1940, in *The Comparative Psychology of Mental Development*, the psychologist Heinz Werner made a study of what he called the 'perceptual and cognitive functioning' of European-Americans and indigenous peoples. And one of the things he found was that indigenous peoples appeared to perceive their surroundings in a much more *real* way. According to him, their perceptions were more 'vivid and sensuous', and he speaks of the 'sensuousness, fullness of detail, the colour and vivacity of image' of which they were aware.[35] The American anthropologist Stanley Diamond reached a similar conclusion, writing that, for indigenous peoples, 'their sense of reality is heightened to the point where it sometimes seems to blaze'.[36]

Cross-cultural psychological studies have supported this too. As the contemporary psychologist Judith Kleinfeld notes in her 1994 essay 'Learning Styles and Culture', tests have shown that while Native Americans, Eskimos and Aboriginal Australians have poorer verbal abilities than European-Americans, they have a much higher level of visual and spatial skills. This presumably explains why Westerners have often been often amazed by the wealth of detail which indigenous peoples can 'read' from their surroundings, apparently able to see, hear and smell phenomena which were imperceptible to them. As Kleinfeld writes of the Eskimos, for example: 'People travelling with [them] remarked on their extraordinary ability to notice and recall visual detail, for example, drawing from memory a map later found to be about as accurate as one made from aerial photographs.'[37] Perhaps this was simply because their heightened perception meant that their surroundings were powerfully real to them.

This is part of the reason why these peoples have such a reverential attitude to nature. This reverence comes partly from their strong sense of psychological attachment to their land, which means that they strongly empathise with natural phenomena, whereas we see them as something 'other' which we have no empathy or responsibility for. But it also comes from their sense that the whole of the natural world is *alive*. Our abusive relationship to nature also stems, I believe, from our familiarised vision of the world. Seen from a Western perspective, nature seems inanimate and unreal, and so we don't have any qualms about using and abusing it. We don't see anything wrong with chopping down trees, paving over fields with concrete, mining the earth for metals, or 'owning' a piece of land. But with their non-automatic perception, indigenous peoples see all things as animate, with their own inner being. All things are pervaded with a spirit force or else inhabited by spirit beings, and so it's impossible for them to conceive of harming or abusing nature.

And if, as all this suggests, indigenous peoples have traditionally a heightened perception of the phenomenal world, then this is another reason why time would pass slowly to them.

Perhaps the most important thing about this for us though, is that it shows that it *is* possible to live in a state of heightened perception and with a less intense ego without suffering any of the terrible effects we looked at in the last chapter (i.e. the effects of drugs and schizophrenia).

In a sense, then, if we want to transcend linear time we should try to change ourselves inside, restructuring our psyche so that we're no longer plagued by constant thought-chatter, and no longer experience the sense of separation and duality which our strong egos generate.

But how can we be expected to change ourselves in this way? After all, you might say, we inherit our 'psyche' from our ancestors in exactly the same way that we inherit our genes and our bodies from them. We wouldn't expect to be able to change ourselves physically – say, to grow extra teeth or develop a more acute sense of smell – so how could we expect to change ourselves psychologically, to somehow develop a heightened perception of the world around us, or to transcend our strong egos?

But the important point here is that the psyche is not fixed or permanent in the same way that the body is. We do have the power to alter it. It is, in fact, very easy for us to 'deautomatise' our perceptions and, to a certain extent, subdue our ego.

6

The Timeless Moment: Higher States of Consciousness and Time

A few years ago my wife and I went to the Isle of Arran, off the west coast of Scotland, for a camping holiday. Unfortunately the elements weren't on our side, and after a few days the rain became so torrential that we had to take refuge in a hotel. It was in a tiny village overlooking the sea, and on our first evening there I covered myself with waterproofs and went for a walk along the coast. The rain was lashing down, the sea was crashing into the rocks, and there were no signs of life apart from a few seagulls flying over the beach. I kept staring at the cliffs, at the white of the sky, at the sea stretching into the distance, and after a few minutes something strange happened. I felt a kind of shift in my vision, and suddenly I had the sense that everything around me was alive. It's not easy to describe in words, but I sensed that the earth was a living being, a sentient being even.

I could sense a force which pervaded everything, and saw that the sky, the sea and the cliffs and even the rain and the air were the manifestations of something greater than themselves. There was a deeper level of unity beneath their superficial separateness, something *behind* them.

At the same time I felt as though I wasn't separate from what I was seeing, that this force which pervaded everything included me as well. I wasn't just observing the scene, I was part of it. My whole being seemed to be overflowing with energy, making me feel exhilarated and euphoric. Nothing seemed to matter, whatever problems and worries I had completely faded away. Above all the world seemed like an incredibly *benign* place, and it seemed somehow massively *right* to be alive in it.

At one point I sat in the middle of a group of rocks and stared at the sea, and I had a sense of being beyond time. I told myself that this was how the world would have been thousands, even millions of years ago – just the sea, the wind, the beach, rain, seagulls circling around; no sign of human life anywhere. I had a sense of the reality of this; for a moment I felt as though I *was* here thousands of years ago, or more precisely, that thousands of years ago was *now*.

Many of us experience 'higher states of consciousness' such as this from time to time: moments in which the world is filled with beauty and meaning and we feel as if we're part of our surroundings rather than just observers; when a sense of inner well-being fills us and we feel as if we're connected to a deeper and truer part of our own self. It's as if we go beyond the limitations of our normal consciousness and experience a more intense vision of reality and a different relationship to the world.

These experiences are much more common that we might first assume. According to one 1987 survey by the researchers Hay and Heald, for example, 16 per cent of British people have sensed 'a sacred presence in nature', while 5 per cent have had a sense of 'the unity of all things'.[1] Similarly, in 1974 the American researcher Robert Greeley asked 1,460 people if they had ever had the experience of being 'very close to a powerful, spiritual force that seemed to lift you out of yourself'. Thirty-five per cent of them said that they had, with 21 per cent saying they had had the experience several times, and 12 per cent that it happened often.[2]

There are two different types of higher states of consciousness. The first are wild, intoxicating ecstatic experiences which can occur when our normal brain or body chemistry is disrupted in some way. Throughout history, religious adepts and mystics have tried to induce higher states of consciousness by fasting, chanting, dancing, controlling their breathing and inflicting pain on themselves – all of which are different ways of interfering with our normal brain/body chemistry and putting our bodies 'out of homeostasis'. Higher states certainly don't always occur in these moments, but they certainly can do, especially in the context of religious rituals or ceremonies. (Drug-induced higher states of consciousness – such as those we looked at in Chapter 4 – belong to this category, since they also disrupt our normal brain/body chemistry.)

The second type of higher states of consciousness are more serene and stable experiences which occur in a more spontaneous way. These can occur as a result of meditating, contact with nature, playing certain sports, having sex, or simply just relaxing. Whereas higher states induced by drugs, fasting or breath control are mainly visionary experiences, making you see the world in a more intense and fresh way, these 'serene' states have a more powerful 'feeling' dimension, with a strong sense of inner stillness and well-being.

There are different levels of the HSCs – as I'll begin to refer to them from now on – and different characteristics reveal themselves at these different levels. At what you could call a low-level HSC, our surroundings 'come alive' to us, they appear more real and beautiful, and the world seems a much more fascinating place than normal. There may also be a sense of meaning and harmony, of a benign atmosphere which seems to radiate from everything. At a medium-level HSC, the sense of being connected to your surroundings is added to this, as the duality between yourself and the 'outside' world fades away. You may also become aware of what Indian philosophy calls *brahman*, or what Native Americans sometimes called the Great Spirit or Life Master, a kind of spirit-energy which pervades everything. This gives us a sense of the oneness of things: along with the separateness between ourselves and the world, the separateness between different things fades away. Everything is a manifestation of this spirit-energy, and so everything is one. (My experience above is probably an example of this level.) This is one aspect of higher states of consciousness which is often interpreted in religious terms. If someone from a monotheistic background has this vision of *brahman* pervading everything, it's likely they will believe that they're experiencing the radiance of God, seeing his divine power radiating through his creation. As the 13th-century Italian mystic Angela da Foligno describes one of her spiritual experiences:

> The eyes of my soul were opened and I beheld the plenitude of God, wherein I did comprehend the whole world ... In all things I saw nothing except divine power,

a power completely indescribable, so that through excess of marvelling my soul cried with a loud voice, saying, 'This whole world is full of God!'[3]

A high-level HSC is what's normally referred to as a 'mystical' experience. At this level duality fades away to the point where we feel that we *are* the universe and everything in it. The whole material world may dissolve into what the Indian mystic Ramakrishna called 'a limitless, shining ocean of consciousness or being'.[4] It's as if we have entered into the absolute essence of reality, the 'ground' of pure spirit which underlies everything and pervades everything. We realise that this consciousness is the source of all things; that the world is its manifestation, that its energy 'pours out' into the world, and that the nature of this energy is bliss and love.

Some scientists see higher states of consciousness as a kind of hallucination or delusion caused by abnormal functioning in the brain. In other words, they see them as a kind of *altered* state of consciousness, like hypnotic trances, dream or hallucinatory states, or states of psychosis. In altered states, the normal ego structure is disrupted in some way, and our normal vision of reality becomes either constricted or completely distorted. In other words, we become *less* conscious than normal. But in higher states the opposite is the case. There's usually no impairment or malfunctioning in our minds, and it's as if we become *more* conscious than normal. Higher states of consciousness bring a powerful sense of revelation. After hallucinations or dreams we're usually aware that what we've experienced is unreal, and that the normal consciousness that we've returned to is fundamentally more authentic and true. But when normal consciousness resumes after a higher state of consciousness, we have the sense that we're experiencing something *less* real and true. Unlike altered states, HSCs always bring a sense of 'so this is the way things *really* are', as if a veil's been pushed aside. We feel that we've 'woken up' in some way, and are now seeing a truer and clearer picture of the world. As the scholar of mysticism R.C. Zaehner writes, for example, a person who has the experience is always 'convinced that what he experiences, so far from being illusory, is ... something far more real than what he experiences normally through his five senses or thinks with his finite mind'.[5] Far from seeing HSCs as something abnormal, we should perhaps see them as a kind of *super*-normal consciousness, which is more valid that our normal state.

What's especially interesting for us though is that higher states of consciousness change our relationship to time. Another of their characteristics is – as we saw with the drug experiences in Chapter 4 – the expanded sense of time they bring, and even the sense that time no longer exists. And as you would expect, this expansion of time varies according to the different levels of the experience. At a lower-level HSC, we may have a similar sense of heightened perception and expanded time to this person's description of how he feels after meditating.

> After a good meditation everything seems more real to me. I look at everything a lot more. I look at the sky and the clouds while I'm walking, at the colours of shops and cars, at billboards and posters, and I feel as if I'm noticing a lot more than I usually do. Everything seems to move slowly and gracefully. I feel like there's no hurry to do anything, because there seems to be more time.

Part of this can be explained in terms of the relationship between time and information. Higher states of consciousness bring about a 'de-automatisation' of perception. The second type of more sedate and stable HSCs seems to occur when there's an intensification of energy inside us, when our minds become extremely relaxed and still and the normal channels through which we expend mental energy (such as thinking, concentrating on tasks and activities and information-processing) are partially closed. This means that our mind no longer has to conserve energy through automatic perception, in the same way that you no longer need to save money after you've received a large cheque. The desensitising mechanism which we discussed in Chapter 2 no longer has to function, and no longer edits out the reality of our perceptions. As a result we experience the is-ness of our surroundings which is normally hidden from us. And this means that we take in much more information from our surroundings than normal, which has the effect of stretching time.

But HSCs also bring an expanded sense of time because they are experiences of ego-transcendence. They usually occur in moments when our minds become much more still and quiet than normal, when all thought ceases and the normal ego structure softens and fades away. Our usual sense of isolation and separateness dissolves, and this is what creates the sense of oneness with the world which often goes with these experiences. In this respect, HCSs are examples of the fifth law of psychological time.

The more our perceptions become 'de-automatised', and the higher the degree of ego-transcendence, the more intense the experience is. And the higher the level of the experience, the more our sense of time expands. At the highest levels of consciousness, there is no time. You could compare this to how speed affects our sense of time. Remember the clock on the spaceship travelling at 87 per cent of the speed of light?

Time on this spaceship would go twice as slow as normal, and if you kept increasing your speed beyond that point it would slow down even further. But what would happen if you could actually reach the speed of light? According to Einstein's equations this is impossible (since mass also increases with speed and at the speed of light your space-ship's mass would be infinite) – but speaking hypothetically, at this point time would actually stop. From the perspective of light itself, there is no time.

And this is what happens as we move through the 'spectrum' of consciousness – from our normal state of consciousness to low-level HSCs through to the highest mystical states. Time keeps slowing down until finally it disappears altogether. One of the best collections and studies of higher states of consciousness is Raynor C. Johnson's *Watcher on the Hills*, originally published in 1959. Johnson includes many descriptions of HCSs from members of the public, together with examples from literature, and one of the defining characteristics of the experience is, he believes, the 'sense of time obscured'. For several of the experiences he collects, people report a sense of going beyond linear time. One person says that the experience 'seemed like the twinkling of an eye; it also seemed like eternity',[6] while another has 'a sense of time arrested and of the feeling that I was looking on a picture as old as the world'.[7]

At the highest-level HSC, time and space dissolve into an ocean of infinite and eternal consciousness; the boundaries between past, present and future melt into oneness. Time is no longer linear, but spatial. Or in the words of the Japanese Zen Buddhist teacher and author D.T. Suzuki: 'In this spiritual world there are no time divisions such as the past, present and future; for they have contracted themselves into a single moment of the present where life quivers in its true sense.'[8] One acquaintance of mine experienced a pow-erful HSC when he was walking through a wood one winter afternoon, looking up at the canopy of trees above him. Suddenly the trees seemed to 'open out into a vision of eternity'.

> Everything had significance, but it meant nothing more (and nothing less) than just *what it was*. At the same time, the inevitability of it all meant that I knew for sure that nothing really mattered, and for an instant I knew also that I was basically immortal, in the sense that death didn't mean anything either.
>
> An eternal state of consciousness seemed to be there in that very simple event … The recognition that took place in that now *was* eternal. The importance of the moment itself somehow defeated death by being outside of time.

This experience echoes the way the great German mystic Meister Eckhart's describes how time dissolves away in mystical experiences.

> This power [the mystical essence of the soul] knows no yesterday or day before, no morrow or day after (for in eternity there is no yesterday or morrow); there is only a present now; the happenings of a thousand years ago, a thousand years to come, are there in the present and the antipodes the same as here.[9]

Similarly, the 19th-century British novelist and mystic Richard Jefferies vividly describes the mystical state he reg-ularly experienced.

> I cannot understand time. It is eternity now. I am in the midst of it. It is about me in the sunshine; I am in it, as

the butterfly floats in the light-laden air. Nothing has to come; it is now ... The years, the centuries, the cycles, are absolutely nothing; it is only a moment since this tumulus was raised, in a thousand years more it will still be only a moment. To the soul there is no past and no future; all is and will be ever, in now. For artificial purposes time is mutually agreed on, but really there is no such thing.[10]

Nature and meditation

In Western philosophy, these experiences – at least at their lower levels – are often referred to as the encounter with the 'sublime'. Philosophers such as Edmund Burke and Kant described the sublime as the transcendent, awe-inspiring quality of nature, the power of some natural scenes to inspire us with joy, wonder or even fear. Hinting at a concept similar to the Indian *brahman*, Kant suggested that the source of the sublime was a 'supersensible substrate' which transcends both the human mind and the natural world. In this way, both my own experience on the Isle of Arran and that of my acquaintance walking through a wood were encounters with the 'sublime'.[11]

The concept of the sublime influenced romantic poets such as Wordsworth, Shelley and Goethe, whose poetry contains many descriptions of higher states of consciousness generated by the beauty and power of natural scenes. In his 'Hymn to Intellectual Beauty', Shelley describes his awareness of a Great Spirit or *brahman* as a 'Spirit of Beauty' pervading the whole of nature, and an 'awful shadow of some unseen power' which 'floats though unseen among us'.[12] Wordsworth's poetry is famous for his sense that the natural world is alive and pervaded with a benign unifying presence. To him, as he writes in *The Prelude*: 'the great mass [of natural forms]/ Lay bedded in some quickening soul, and all/ That I beheld respired with inward meaning.'[13]

And nature is, it's true, one of the most powerful triggers of HSCs, due to its powerful 'energy-intensifying' effect. For sedate and stable higher states, the important thing is simply to be in a situation which helps to generate a higher than usual concentration of mental energy inside you, usually by conserving it. And when you're walking in the countryside on your own, you conserve energy by not being active or communicating with anyone, by having quietness and stillness around you (so that you won't expend much energy through processing information) and by (hopefully) having a lower level of thought activity (because you're ideally giving your attention to the beauty around you rather than to your thoughts). As a result, you build up a higher intensification of energy inside you. This brings about the heightened perception, the sense of connection to your surroundings and the inner well-being which Wordsworth frequently describes, and which I experienced on the Isle of Arran.

Probably the most reliable way of generating higher states, however, is through the practice of meditation. This is because meditation is the most effective way of quickly intensifying and stilling our mental energy. When we meditate we close our eyes and are in silence, so that we drastically reduce the amount of information we process. We don't concentrate on anything apart from the focusing device we're using (a mantra, our breathing, a candle flame, etc.), and hopefully – if we have a good meditation – the thought-chatter which normally runs through our minds will fade away. As a result, with its 'outflow' drastically reduced, we build up a high concentration of energy inside us, which brings about the de-automatised perception, the sense of inner well-being and other characteristics of HSCs.

Meditation certainly doesn't always bring higher states of consciousness though. Three different laws of psychological time can operate during it, depending on the meditator's state of mind and his or her ability to concentrate. Anybody who has sat down to meditate and had difficulty concentrating on their mantra or breathing – especially if it's a group meditation and you feel that you can't get up and leave the room for fear of disturbing the others – will have experienced the effects of the third law, when time goes slowly due to non-absorption. Your legs start to ache, different parts of your body start to itch, your mind jumps chaotically from one random thought to the next, and you can't imagine how you're going to endure this for another 15 minutes or so. And by the time the meditation leader asks everyone to open their eyes you feel like you've been sitting there for an eternity, and are shocked that it was only 20 minutes. And this is, of course, because of the massive amount of thought-information that runs through your mind, stretching time in the same way as when you're waiting for a bus or listening to a boring lecture.

However, if you can concentrate fairly well and become absorbed, the fourth law will operate and time will go quickly. In Chapter 3 I mentioned that the people doing meditation exercises on my courses usually feel that less time has passed than actually has. The American psychologist Arthur J. Deikman found this too, when he conducted experimental meditation sessions, in which he simply asked volunteers to concentrate their attention on a blue vase. He told them to try to see the vase as it existed in itself, without connection to other things, and not to be distracted by thoughts. The volunteers did this for ten half-hour sessions spread over a month, and reported a number of perceptual changes. They said that the vase became more beautiful and vivid to them, even to the point of seeming alive, and that they lost their sense of being separate from it. As one person told Deikman: 'I really began to feel as though the blue and I were merging or that the vase and I were one. It was as though everything was sort of merging.' And in terms of time, the volunteers generally felt that, in Deikman's words, 'less time had elapsed than was recorded on the clock'. However, this was only if they could concentrate well. If they kept losing concentration, they experienced the fourth law, finding that the periods seemed to drag frustratingly, and to last much longer than 'clock time'.[14]

However, there's a deeper level of meditation than this – when the ego falls into abeyance – at which the fifth law starts to operate. If you sit down to meditate and feel especially relaxed and alert, you might find that after a certain amount of concentration your mind becomes completely quiet. The normal flow of thought-chatter fades away, and there's nothing between you and the pure consciousness of your being. You feel that your normal ego has dissolved away and you've become a different 'I', the deeper and more expansive self which (as I described in Chapter 4) still remains when thinking stops. You don't feel the need to think, do or be anything but simply rest within your own being, experiencing the emptiness and freedom of pure consciousness.

At this point you may feel that you've transcended time, that you're experiencing a wholly different realm of being where concepts of the past, present and future cease to have any meaning. You experience a similar egoless state to sportspeople in the Zone, or people who have 'time-slowing' experience in accident or emergency situations.

The ego which creates our sense of time fades away, and so time fades away too.

If you have a good meditation, the effects usually stay with you for hours afterwards. I usually find that, if my early morning meditation goes well, it puts me into a very mild HSC until at least lunchtime, if not longer. I feel a sense of inner well-being, and my perceptions are at least partly de-automatised, so that I'm more awake to the is-ness of my surroundings. It's like seeing the world through the eyes of a child: objects seem more real, and I find myself noticing things I wouldn't normally pay attention to, such as the patterns on cups or shirts or of the sunlight shining on to the kitchen floor. I seem to become more sensitive to beauty, and find myself staring at the sky or at trees or flowers. (Deikman's volunteers experienced this too. After a session one of them said that everything he saw from the window seemed to be 'clamoring for his attention', while another was amazed at the beauty he could see everywhere, which he described as 'perception filled with light and movement, both of which are very pleasurable. Nobody knows what a nice day it is except me', he concluded.)[15] And like the person I quoted earlier (who said that after meditation 'everything seems to move slowly and gracefully'), I feel as though I can glide easily through the day, giving complete attention to everyone I meet and everything I need to do. I don't feel the need to rush anything, partly because I feel so relaxed, but possibly also because I have a slightly expanded sense of time.

High-level HSCs induced by meditation are less common, but do occur sometimes. Here a friend of mine – the author and publisher Colin Stanley – describes a high-level HSC he experienced while meditating. The experience was triggered by staring at a candle flame, and lasted for several hours afterwards.

> The image of the flame became concentrated, a pinprick of light. Then it exploded, filling my head with a brilliant white light. I felt a corresponding surge of joy and security (as if nothing could ever harm me) and a feeling of timelessness which (this sounds like a contradiction) seemed to contain all of time.
>
> For several hours I felt as if I was above myself, floating in the air yet able to feel my body precisely, as I had never felt it before. I went out into the grounds of the house and was aware of every step I took, every movement around me and every sound carried on the air. I felt, for the first time ever, as if I was in total control of my mind and body. The rest of the day was like an eternity. I seemed to move around, not necessarily slowly, but incredibly precisely.

'Positive' wakefulness

We've already said that the key to slowing down time in our lives – and making ourselves less subject to time in general – is to find a way of making ourselves more 'awake' to our experience, so that we take in more perceptual information, and of subduing or weakening our normal ego. We've also seen that there's obviously no point recommending a constant diet of LSD as a way of transcending time, because of the psychological damage this would cause.

But the serene and stable HSCs we've been looking at don't have any bad side effects. After long periods of regular LSD trips or during schizophrenia the ego breaks down completely, so that you can't think rationally, make decisions or concentrate, you don't have a clear sense of reality (so you aren't sure whether your fantasies and daydreams are real or not), and may suffer from sensory distortions. But in these serene higher states, the ego isn't destroyed – in fact in a way it becomes healthier than normal, since with its crazy thought-chatter quietened, it becomes more still and more stable. If I have a low-level HSC after a good meditation, for example, I always find that I'm much better at concentrating and thinking logically, precisely because my mind isn't as disturbed by chaotic thoughts. The important point is that, in these higher states, the ego isn't dismantled or damaged, as it is with LSD experiences. It stays intact as a structure, but it's quiet and still, like a house that remains standing even though it isn't being lived in.

The question we need to ask, then, is: is it possible to live in a permanently higher state of consciousness, at least at a lower level? I would argue that indigenous peoples like the Trobriander Islanders, the Nuer of Africa and Aboriginal Australians already did – and perhaps still do – this, at least to an extent. The vision of the world we have in higher states is their normal, everyday vision. In fact, to be more specific, their normal vision of the world seems to correspond roughly to low- to middle-level HSCs – with a heightened awareness of reality; an awareness of a spirit-force pervading the whole world; and a sense of the interconnecteness of phenomena and of their own connection to the whole of nature. For us though, this would mean self-transformation, a complete reorganisation of our psyche.

However, I believe that it's possible to transform ourselves in this way, and that the key to it is the practice of meditation, together with what Buddhism calls mindfulness – the act of paying complete conscious attention to everything we experience and do. Meditation is an extremely powerful tool of transformation, not just because it's the most reliable way of generating higher states, but because it has the long-term effect of permanently transforming our psyche, slowly leading to a permanently higher state, if only a low-level one.

Before we look into this potential transformation in detail though, we need to investigate a theoretical question which our examination of HSCs has given rise to: the possibility that time is really just an illusion.

7

The Illusion of Time

How can time be an illusion? If it doesn't really exist, why is that we get old and die? Why does the day start at dawn, last for a certain amount of time while the sun moves through the sky, and finish at dusk? Why do songs on a CD have a beginning, last for, say, 3 minutes 15 seconds and then finish?

The idea that time flows in this way seems completely obvious, and is what you could call the common sense view of time. According to Stephen Hawking, there are three pieces of evidence for this view. Firstly, there's our personal sense of time – our sense of ageing and our sense that events and processes last a certain period of time. Secondly, there's the 'time' of the universe – from the Big Bang through its expansion in space towards its final death in a Big Crunch. And finally, there's the time of the second law of thermodynamics – the fundamental law of physics which says that all processes in the universe gradually lose energy and run down, moving from a state of order to one of entropy or disorder.[1]

But despite this, perhaps the idea that time is *not* linear, that in some way the past, present and future exist simultaneously, is not as nonsensical as it seems. We've just seen that this is how time appears at the highest levels of consciousness, and the fact that our sense of time is so closely connected to the ego supports this view too. As we saw in Chapter 4, time seems to slow down or stop in non-egoic states such as accident and emergency situations, states of intense concentration, drug experiences, mental illness and hypnosis. As I noted then, this seems to suggest that time doesn't exist *outside* the ego, but is created *by* it. And after all, young babies appear to live in a timeless realm before their ego develops, and slowly 'fall into' time as they gain a sense of self.

Timelessness in science

The idea that time has no absolute and objective reality is, of course, suggested by Einstein's Theory of Relativity. It could be that all the 'time distortions' which the theory suggests are simply distortions of something which is objectively real, in the same way that, say, five different people who witness the same accident from different vantage points give subjectively distorted versions of an event which actually happens. But the distortions are so numerous and so fundamental that it perhaps makes more sense to assume that there's actually nothing there to be distorted in the first place. As well as showing that time is relative to speed and gravity, Einstein showed that there's no set order in which events occur. From your vantage point two events might seem to happen at the same time, but to someone who's moving at high speed there might be a time lag between them. Or there might be two events which happen one after the other, which a person travelling at a high speed sees in reverse order to you. Einstein also showed that the speed of time varies according to how 'curved' a particular area of the universe is. Space-time can be curved by massive bodies with a powerful gravitational pull, and when this happens, the speed of time slows down. The greater the curvature, the slower time passes. And in the case of black holes, when massive stars collapse in on themselves and their gravitational pull is so great that even light can't escape from them, this means that time stops altogether.

In other words, time is something which is created *by* things, not something which existed before them or which could exist without them. In the words of the science writer John Boslough: 'Time is a local phenomenon throughout the universe, quite literally a creation of the gravitational field of each celestial body and unique to any locale in the cosmos.'[2]

Modern physicists also tell us that we should think of the universe as existing in four-dimensional space-time, and that in these terms the idea that time flows makes no sense. Rather than flowing, time is just *there*, in the same static way that space is. In other words, the whole of the past and the whole of the future are here now, existing side by side with the present. In four dimensional space-time, the concepts of past, present and future are meaningless. According to Oxford Professor of Mathematics Roger Penrose: 'The way in which time is treated in physics is not essentially different from the way in which *space* is treated ... We just have a static-looking fixed 'space-time' in which the events of our universe are laid out!'[3] The idea that time passes, or that there is a past and a future divided by a present, has never been verified by any physical experiments. As another modern physicist, Paul Davies, notes: 'As soon as the objective world of reality is considered, the passage of time disappears like a ghost in the night.'[4]

And even if, for argument's sake, we accept the idea that time flows, there's nothing in physics to say that it has to flow in a forward direction. It makes no difference to the laws of nature *which* direction time goes in; they work just as well whether it's going backwards or forwards. After putting forward the evidence for the common sense view of time, Stephen Hawking presents the case for the non-linear view, noting that 'the laws of science are unchanged' when particles are changed to anti-particles, when their 'left' and 'right' sides are swapped, and when their motion is reversed (in other words, when they move backwards in time).[5] The space-time diagrams which physicists use to picture the interactions of particles can be interpreted either as showing positrons moving forward in time, or electrons moving *backwards* in time. Both interpretations are equally valid, and identical from a mathematical point of view.

And even in the macrocosmic world, some physicists believe that it may be possible to travel back (and forward) in time, through wormholes. These occur, at least in theory (since nobody has actually found evidence of their existence), when the fabric of space is ripped by massive bodies like black holes, creating tunnels between different times and places. Although Einstein's equations predicted that it's impossible to travel faster than light, some physicists believe that particles called tachyons may exist, which have no mass and so can reach – and go beyond – the speed of light. And this could mean jumping forwards or backwards in time too.

Einstein's Theory of Relativity was part of the early-20th-century revolution in physics, when the solid, deterministic world of Newtonian physics dissolved into a nebulous realm where there are no absolutes and nothing is objectively true. The other part of this revolution was the development of quantum physics, the study of how subatomic particles – that is, particles smaller than the atom – behave. In this subatomic world, all of the laws which govern the macrocosmic world break down. Particles can vanish and reappear seemingly at random, they can affect each other's movements even if they're miles apart, and a human observer can change the way they behave. And on top of this, the issue of what a particle *is*, is also far from straightforward. A 'particle' may appear to be a hard, solid entity like a tiny billiard ball, but in reality it's a whirl of energy which manifests itself as a particle at that particular moment. At other moments – and this is one aspect which can be affected by a human observer – particles can appear as waves.

As we'd perhaps expect, in this subatomic world, linear time breaks down too. Our normal concepts of the future and the past have no meaning. The physicist Werner Heisenberg was the originator of the famous 'uncertainty principle', which shows that it's impossible to measure both the speed *and* the position of a subatomic particle at the same time. And with reference to time, he wrote that: 'In elementary particles space and time are strangely blurred in such a way that one can no longer define correctly the concepts of earlier and later … When experimenting within very small space-time domains, one should be aware that process could run in a timely reverse order.'[6] In other words, in the quantum world the normal forward flow of events sometimes runs backwards. When an explosion takes place inside an atomic nucleus, sometimes it is *followed* by the cause of the explosion, rather than the cause happening first.

It seems impossible to explain any of this without accepting that our sense of the sequence of time – and therefore time itself – is an illusion. This is exactly the view which one eminent British physicist, Julian Barbour, has recently put forward. In his book *The End of Time*, Barbour suggests that in reality the universe exists in a timeless state. According to him, our lives consist of 'snapshots' of experience – isolated and separate moments, like the frames of a film. It's because these 'snapshots' appear in sequence to us that we have the experience of time passing. Each 'frame' is a particular arrangement – or 'configuration' – of all of the particles in the universe. At one particular moment all particles are 'configured' in such a way that, for example, I'm sitting at my writing table with my hands raised a couple of centimetres above my computer keyboard and my fingers poised to press the buttons; meanwhile the branches of the trees outside my window are poised at a certain angle after being blown by the wind, the aeroplane high above me has reached a certain point in its journey, the girls in the school at the back of my house are frozen in mid-motion in their lessons, and so on. All of these things – me, the trees, the aeroplane, the girls – are made up of particles, and all of this is just one possible way in which all the particles in the universe could be arranged at any possible time. In the *next* moment, the particles of the universe are arranged in such a way that my hands press the computer keyboard, the aeroplane moves further in the same direction, and the girls move on to the next point in their lesson. These 'nows' don't have any time involved in them – in essence, they are static. But because they appear in sequence we have the impression that time is 'flowing'.

Barbour asks us to imagine that in reality the universe is a kind of static ocean in which all the possible arrangements of all particles exist side by side with one another. He calls this 'Platonia' (hinting at a plateau and Plato's world of ideals), which he describes as 'the totality of possible nows'. As we live our lives, different nows are 'switched on', or 'chosen' out of all the possibilities. Hypothetically, we could live in a different universe that selected nows which aren't in sequence, so that the now of me with my fingers poised over the computer keyboard might be followed by a moment from 10,000 years ago or a million years in the 'future'. But fortunately for us, we live in a universe which selects the nows – or possible arrangements of particles – which follow on from one another, and so gives us linear time. In the analogy Barbour uses, the pack of cards (representing all the possible nows) is dealt out to us in sequence, whereas in another universe they might be handed out completely randomly.[7]

Perhaps, however, the view of time of post-Newtonian physics was best summarised by Einstein after the death of his friend Michele Besso in 1955, in a letter to Besso's son and sister:

> He has preceded me briefly in bidding farewell to this strange world. This signifies nothing. For us believing physicists the distinction between past, present and future is only an illusion, if a stubborn one.[8]

Precognition

Another reason for doubting the straightforward linear view of time is that, under certain circumstances, the veil which prevents us from being aware of the presence of the past and future seems to fall away, enabling us to catch glimpses of past or future events.

Some sceptics and materialists may doubt that this is possible, but the existence of precognition (and *retro*cognition, the experience of reliving the past) makes sense in terms of the picture of time we've been developing, and in terms of modern physics. After noting how the normal sequence of events is sometimes reversed in the quantum world, the physicist Pascual Jordan (who was one of the founders of quantum physics alongside Heisenberg and Max Born) remarked that: 'This has enormous implications for psychology and parapsychology, since such reversals of the cause-and-effect sequence are proved possible and philosophically valid.'[9] If linear time is only a creation of the mind, if the future already exists and past still exists, why shouldn't we be able – in certain circumstances and certain states of consciousness – to experience them?

Anybody who approaches the evidence for precognition with an open mind will find it very difficult to dismiss. As far back as the 1930s, the researchers J.B. and Louisa Rhine found that volunteers could guess which cards were going to be taken at random from a pack with a success rate 3 million times higher than chance.[10] More recently, the physicist Helmut Schmidt found that volunteers could predict random subatomic events, obtaining results – in over 60,000 trials – which were a billion to one against chance. Over a 25-year period, the Dutch psychic Croiset has consistently demonstrated the ability to accurately describe people (including their occupation and past experiences) who are going to sit in a particular chair in an auditorium at a particular event taking place in the future.[11] In all, between 1935 and 1987 almost 2 million individual trials took place to test for precognition, involving 50,000 people. These were reported in 113 scientific articles, and overall they showed incontrovertible evidence for the reality of the phenomenon, with odds against chance of 10^{24} to one.[12]

In addition to these scientific studies, however, there's massive anecdotal evidence for precognition. Scientists are often suspicious of anecdotes, preferring to stick to hard facts which they can verify (or not) through experiments. But surely there are some cases when anecdotal evidence is so widespread and persuasive that it has to be taken seriously? Many reports of precognition are so strikingly accurate that to refuse to consider them because they are 'just' anecdotal seems short-sighted.

Most instances of precognition take place in dreams. One of the most famous and the best authenticated cases is that of a man called John Godley, who dreamt the names of winning racehorses. One night in 1946, when he was a student at Oxford University, he dreamt he was reading a list of horse race winners in a newspaper and saw the names Bindle and Juladdin. The next morning he told a friend and they checked a newspaper, finding that two horses with those names were running that day. He and a group of friends decided to risk a bet; both horses won and they won large sums of money. This happened to Godley another two times over the next three months, and several times over the following years. In addition, he occasionally dreamt of horses' names which did run in races (which is already precognition in itself) but which didn't win.[13]

Whenever I run courses I always ask my students whether they've had any experiences of precognition, and they always supply me with examples which seem far too precise and striking to be explained away as coincidence. On one of my most recent courses a woman told me about a nightmare in which she answered the phone to her mother, who told her the terrible news that her sister had died in a car crash. She awoke with a feeling of dread and phoned her sister that morning, telling her not to go out in the car today, without actually telling her about the phone call. But her sister disregarded her advice, saying she had no choice because she had to travel to a different town for her job. And that evening my student's phone rang – it was her mother, telling her exactly the news she was dreading.

Just a few days before my writing this, my wife – who works as a podiatrist – had a dream in which she accidentally cut off the toe of a healthy patient.

> I was pleading with a surgeon to sew it back on, knowing that if he did it quickly enough the blood supply would return. But the surgeon wasn't listening and the toe stayed off ...
>
> I thought it was a strange dream because cutting people's toes off isn't part of my job – I'd only ever done it once before, about ten years ago. But that day at work, I was treating a guy with a gangrenous toe, who was fed up of waiting for it to drop off. So I decided to cut it off. And it hit me just a few minutes later – it seemed so bizarre that I should cut someone's toe off for the first time in ten years, the day after dreaming about it.

However, most examples of precognition are more mundane. Typically, a person might see a particular scene – or series of scenes – in their dreams, which they then experience the following day. One of my students told me about a dream in which he'd been in the hall of an unfamiliar house. He saw one of his friends walking downstairs with a strange woman, carrying a silver object which was reflecting light brightly. Despite its unremarkable content, the dream seemed unusually vivid, and he told his wife about it over breakfast. The following night, he and his friends went to a pub and ended up being invited back to a house party. At one point he was standing in the hall, and his friend came downstairs with a woman who he'd paired off with at the pub. He recognised the scene of his dream immediately – and realised that the shiny object was a CD, which his friend was bringing downstairs to listen to.

Premonitions

In the 1950s, the American psychic researcher W.E. Cox made a study of railway accidents in the United States between 1950 and 1954. He looked at passenger figures for other trains on the day of the accident, for the same train on the previous seven days, and for the same train on the same day of the week over the previous four weeks. In every case, he found that significantly fewer people than normal travelled on the trains which had the accidents. And the reason for this, he suggested, was that some of the train's potential passengers had premonitions of the accident and decided not to make their journeys.[14]

This explanation may seem far-fetched, but there are many well-authenticated cases of premonitions before major accidents and disasters. One of the biggest local catastrophes in Britain's recent history occurred in 1966, in the Welsh mining village of Aberfan, when a massive coal slip came down a mountainside and engulfed houses and a school, resulting in the death of 144 people, including 128 children. One psychiatrist who worked with the villagers after the disaster, J.C. Barker, found that 76 people there had had premonitions of the accident. Some of these were just feelings of intense anxiety – a feeling that something terrible was going to happen – but others were actual visions of the events in dreams. In all, 36 people 'saw' the accident in their dreams, and in some cases the visions were so vivid that they woke up in fright.[15] It's well known that premonitions caused some people to cancel bookings on the *Titanic*, while more recently, the scientist and psychic investigator Rupert Sheldrake has found that many people had premonitions of the 9/11 disaster. Sheldrake asked for information about premonitions and dreams through posters and a newspaper advert, and received 57 accounts which he felt were specific enough to be relevant. About two-thirds of them were visions of the events in dreams, the others just premonitions of danger, and most people were so frightened by the dreams or their fears that they told people close to them about them before 9/11. Most of the dreams took place either the night before the disaster (about a third) or during the preceding week (another third). One person described his dream as follows.

> Some five days prior to the 9/11 disaster, I had an unusually vivid dream. I dreamt that I was a passenger on a commercial jet … I remember a pervasive sense of dread. The passengers and I were deeply concerned about the flight path we were taking; we were flying low over Manhattan's buildings … I was very frightened about how close we were to the buildings. Many of the passengers were vocal and shared my concern. It was clear that we were flying directly over the southern tip of the island. Then there was tremendous impact and I woke up.[16]

One of the most unusual examples of this phenomenon is the British 'psychic artist' David Mandell. Twenty years ago Mandell, a retired lecturer, began to have vivid dreams which he believed contained visions of the future. Since he was an artist, he started to paint scenes from the dreams. Then he would go to his local bank to have them photographed underneath an electronic clock, showing the date and the year. Eventually, he realised that he was seeing visions of future world disasters. For example, one of his paintings showed sixteen blood-covered children and a teacher being attacked by a man. Mandell believes this was a vision of his premonition of the attack at Dunblane primary school in Scotland in 1996, when a man walked into the school gym and shot dead sixteen children and their teacher. Another painting showed a gas attack at an underground station, and in his notes accompanying the picture, Mandell mentioned that it would take place in Tokyo, and referred to a 'secret group' who came from the 'hills'. This clearly relates to the attack by members of the religious cult Aum Shinrikyo in 1995, who released sarin gas into underground trains in Tokyo, killing twelve people. The group did live in the hills outside Tokyo.

Mandell also painted a picture showing a Concorde aeroplane crashing at an airport, anticipating the crash in Paris in 2000. Mandell included a French flag in the picture, indicating that he was aware of the location. But most strikingly of all, on 11 September 1996, he photographed a painting which showed two burning towers falling into one another and also included an outline of the head of the Statue of Liberty and the silhouette of an aircraft flying downwards.

The similarities in these cases – particularly with his paintings of the underground attack and the twin towers – are so striking that it's difficult to see how they could be anything other than genuine precognitive visions. The specific details are highly impressive and surely too precise to be explained in terms of coincidence. There has been some experimental verification of Mandell's claims too. He passed a lie detector test, showing that he was not being intentionally fraudulent, and the photographs negatives were examined by forensic experts, who found no signs of tampering. In an experiment devised by the sceptic investigator Chris French, twenty people were shown Mandell's paintings and given his own interpretations of them together with alternative interpretations. They were asked to choose which best suited the paintings – and in 31 out of 40 pictures, all twenty people chose Mandell's interpretation. French himself said this result was statistically very significant, so that Mandell's claims 'merit serious consideration'.[17]

Premonitions of disasters may not simply be a human phenomenon either. There are many cases of animals having 'premonitions' of disasters like earthquakes and avalanches – for example, cats running away or hiding, dogs howling and barking for hours, birds escaping their cages and behaving unusually. Here, for example, a villager describes how strangely his animals behaved in the hours leading up to a serious earthquake in Greece in 1999.

> Nearly all our birds – chickens, pheasants, partridges, ducks and geese – showed enough nervousness … The geese made a lot of noise and ran hither and thither seemingly aimlessly and individually rather than in their usual unified manner. I thought there must be a fox about. This behaviour became progressively more intense and by the time of the quake they were quite panic-stricken, running about their enclosure as if they wanted to escape. About five minutes before the earthquake my dogs began to howl in a way that I had never heard before.[18]

You could argue that these 'powers' are simply the result of the more acute senses which animals have. Perhaps the animals sensed the vibrations of the earthquake while they were still too faint for humans to detect, or perhaps they were sensitive to changes in weather conditions. However, there are many cases of animals giving warning of imminent events when they couldn't possibly have sensed them

In the Second World War, many families relied on dogs to give them warnings of air raids. The dogs gave these warnings while the planes were still hundreds of miles away, when it would have been impossible to hear them. Some dogs in London even gave warning of V2 supersonic rockets, which could not possibly have been heard by them.[19]

A few years ago, when I was living in Singapore, I witnessed a similar phenomenon. We lived close to a street on which the ladyboys – men dressed as women, some of whom had had operations to make themselves more feminine – offered their wares as prostitutes. They were bizarre characters, and sometimes my wife and I liked to just sit in the bar opposite and watch them strutting around. What they were doing was illegal, of course, but they were never caught by the police because of the amazing dog which they had somehow trained to warn them of approaching police cars. The dog spent all its time lying quietly on a step, but every so often it would suddenly become massively agitated, barking loudly and dashing around. The ladyboys took this as their signal to hide, and without fail, a minute or so later a police car would arrive on the scene. I saw this several times, and heard a lot of different people remark on it. The obvious explanation is that the dog's extra sensitive hearing was able to pick up the police car while it was some distance away, but the cars never used sirens, so it's difficult to see what it could actually have heard. Perhaps the police cars had a particular engine sound, you might say, but there was so much traffic on the roads, and the police cars weren't an unusual type, so it's difficult to see how they could have been trained to listen for them. Since the dog usually reacted at least a minute before the car arrived, it's hard to come up with any rational explanation for this ability.

In addition, many people who are epileptic believe that their dogs can give them warning of their fits, enabling them to prepare for them. A study of 21 dogs with this ability showed that before a seizure began the dogs all became anxious and restless. They barked and whined, jumped up and nuzzled their owners, encouraging them to sit or lie down. The study showed that the dogs were always accurate. Of course, it's possible that the dogs might be reacting to sensory cues in these cases too, but significantly, some dogs can still give warnings of fits even if they are in another room. In the same way, there are cases of dogs anticipating heart attacks and diabetic comas. In a recent article in the *British Medical Journal*, a woman told how she relies on her three-year-old golden retriever to tell her when she's going to have hypoglycaemic attacks (due to her diabetes). The writer notes that on the day of the attack the dog 'paces up and down and puts his head on her lap; during nocturnal episodes, he barks and scrabbles against the bedroom door. Nat only settles once her hypoglycaemia has been corrected.'[20]

Of course, all of these are just accounts and it could be that in some cases they are distorted by exaggeration or embellishment. Perhaps in some cases – although I believe this is unlikely – the reports are completely fraudulent. But the combination of the scientific evidence from tests for precognition and the striking anecdotal evidence makes it hard to doubt that it's a real phenomenon.

And it's worth reiterating that none of these experiences should seem at all surprising in terms of the view of time this book has put forward. Experiences like precognition are often seen as antithetical to the scientific worldview, but in actual fact they are fully compatible with the non-linear view of time suggested by modern physics. Since there's no passage of time in physics, and no distinction between the past, present and future, it's feasible that – in special circumstances – we should be able to glimpse or anticipate the future. (Part of the problem here, I believe, is that many sceptics and materialists don't fully grasp the implications of modern physics, particularly of quantum physics.) The future already exists, but our normal egoic consciousness confines us to a linear view of time, so that normally we don't have access to it.

Indeed, there's certainly a connection between these experiences and the experiences of ego abeyance which we looked at in Chapters 3 and 5. Although sometimes they seem to come to us in ordinary states of consciousness, most precognitive experiences occur in moments when the ego is either absent or relatively subdued – in dreams, or moments of deep relaxation or absorption, when the conscious mind is inactive. And they usually *don't* occur – except in cases of especially gifted psychics like Croiset – when we try to *make* them happen; that is, when we make a conscious effort to generate them. Our normal egoic state of mind – when our thoughts are chattering away – seems to stand in the way of psychic powers. I often find myself opening a dictionary or book at exactly the right page I'm looking for. This happens to me so often that I'm fairly certain that it's not just coincidental, and that some kind of higher function of my mind is making it happen. But it *only* happens when I do it without thinking, usually when I'm in a state of absorption and reach for the book and open it completely automatically. Whenever I'm aware of what I'm doing, and make a conscious effort to open it at the right page, it never happens. This is probably one of the reasons why, although there have been many successful laboratory demonstrations, some psychics find it difficult to reproduce their powers under 'scientific' conditions.

Even the unusual powers which some animals appear to possess make sense in these terms. One controversial theory is that what we know as 'psychic powers' such as telepathy or precognition are in some sense natural to all species. After all, even insects like termites work together communally (building skyscrapers, hollowing out trees or moving soil) with an amazing degree of organisation and precision which suggests that a form of telepathy or 'group mind' may be operating. In human beings, these powers don't often manifest themselves because our busy thinking egos get in the way. However, animals like dogs don't have thinking egos in the same sense that we do, and perhaps, as a result, their natural psychic abilities are free to manifest themselves.

Retrocognition

If the future already exists, then the past *still* exists, and just as it's possible for us to glimpse the future, it should also be possible for us to re-experience the past. Retrocognition, as this is known, appears to be much less common than precognition. However, there are a number of well-verified reports of the phenomenon.

One famous instance took place in 1901, when two English women – Annie Moberly and Eleanor Jourdain – were walking through a park at the palace of Versailles near Paris. They started to feel depressed for no apparent reason and were surprised to see a number of people wearing 18th-century costumes. They were given directions by two gardeners and warned by another person not to go down a certain path; then they passed a woman who was painting. A few months later one of the women returned to Versailles and had the same 'eerie feeling', as if she had 'crossed a line and was suddenly in a circle of influence'. This time she saw two strangely dressed workers loading a cart, but when she glanced back a second later they'd disappeared. She heard voices and the sound of dresses rustling but couldn't see anybody, even though she was in a wide-open space.

Three years later the women returned to the park and found that everything was completely different. The park they remembered had had a bridge, trees, a ravine, and a kiosk, but now these had disappeared. After researching the history and geography of Versailles, they concluded that they had slipped into a 'timehole' and seen the park as it was over a hundred years ago, shortly before the French revolution in 1789. This explains the atmosphere of depression and fear they experienced. Moberly and Jourdain wrote an account of their experience – published as *An Adventure* in 1911 – and after its publication, many people came forward with similar experiences in the same park. For example, a family who lived in a flat overlooking Versailles said that they had seen the painting woman in the park twice, and seen other people in 18th-century clothes. A French couple had visited the park the year before Moberly and Jourdain and saw the same gardeners and ploughmen mentioned in their account.[21]

Another instance of retrocognition occurred in 1957, when three young Navy trainees walked into a village called Kersey in Suffolk as a part of a training exercise. It was a beautiful Sunday morning, when you'd expect to see people heading to church and talking in the streets, but the village was strangely deserted. There were none of the normal signs of 1950s life – no cars, telephone wires or radio aerials. It was also as if the seasons had changed – although it was November and the fields they'd just seen had an autumn colour, in the village it looked as though it was suddenly early summer, with the trees full of leaves. As one of the three men, a Mr William Laing, told the parapsychological investigator Andrew MacKenzie:

> There was a butcher's shop, the only commercial building we saw in the village ... There were no tables or counters, just two or three whole oxen carcasses which had been skinned and in places were quite green with age ... Who would believe in 1957 that the health authorities would allow such conditions ...
>
> When one walks past homes, especially in a neighbourhood like that, one sees evidence of wallpaper or flower vases in windows etc., but in the laneway the windows of every house reflected back darkly ... We looked through windows, and I clearly remember a smallish room with a rear window. There was no furniture inside, no curtains, the white paintwork was dingy, certainly not of modern quality.[22]

MacKenzie's investigations led him to suggest that the men had experienced the village as it was more than 600 years earlier, most probably in 1349, after it had been ravaged by the Black Death.

These are isolated incidents of retrocognition, but there are also cases of individuals who have a recurring ability to experience past events, in the same way that some psychics have a recurring ability to glimpse the future. The historian Arnold Toynbee had many experiences of retrocognition, in which he would witness historically significant events. He would, in his words, be 'carried down in a "Time pocket"' and have the 'experience of a communion on the mundane plane with persons and events from which, in his usual state of consciousness, he is sundered by a great gulf of Time and Space'. In these moments he would be 'transformed in a flash from a remote spectator into an immediate participant, as the dry bones take flesh and quicken into life'.[23]

In 1921, for example, he was in the theatre of the ancient city of Ephesus in Turkey, when suddenly he found himself witnessing a scene that had taken place there almost 2,000 years ago. The empty theatre suddenly 'peopled itself with a tumultuous throng as the breath came into the dead and they lived and stood up upon their feet'. Two Christian missionaries, Gaius and Aristarchus, had been taken to the theatre, after coming to the town to denounce the citizens' pagan worship of the goddess Diana. Toynbee hears the crowd chanting 'Great is Diana!' and wonders if the two men are going to be beaten to death. The town clerk intervenes, trying to reason with the crowd, but at that moment the scene fades, and Toynbee is 'carried up again instantaneously to the current surface of the Time-stream from an abyss, nineteen centuries deep, into which the impact of the sight of the theatre at Ephesus had plunged him'.[24]

Some of Toynbee's experiences occurred at the scenes of historically significant events like this one, but others occurred simply when he was reading. He describes how, as a student at Oxford in 1911, he was reading the historian Livy's account of the war of Rome and its allies between 90–80 BC. As he read a passage describing how one Roman leader committed suicide after being betrayed by his wife, Toynbee was 'transported, in a flash, across the gulf of Time and Space from Oxford in A.D. 1911 to Teanum in 80 B.C.' He describes the scene he witnessed in great detail; he's in a backyard at night, watching as the man creeps to his home, sure that he's going to find safety with his wife. But then:

> ... in answer to his low call, a woman's head appears at the window, and one short colloquy informs him that his wife is as heartless as his comrades-in-arms. In an instant, the blade rasps in the scabbard, the body falls with a thud, and the splashing blood irrevocably seals the traitor-wife's infamy. Already the beat of the avenging Furies' wings can be heard in the air as the twentieth-century eye-witness is caught up again and replaced in a trice in his normal locus in Time and Space.[25]

Perhaps we sometimes experience a kind of retrocognition in our own lives too. Occasionally the past becomes incredibly real to us, so real that we feel that in some way it still exists. For example, you might hear a song on the radio that you haven't heard for years, which you associate with a particular time in your life. As you listen, the song takes you back to the situation you were in when you heard it before. Suddenly it's as if a door opens. You can sense the atmosphere of the flat you were living in at the time, the sights and smells and sounds of the area you lived in, and you re-experience yourself as you were at that time, feel the same feelings and think the same thoughts. You can sense it all so vividly that it seems to be more than just memory. Or this might happen when you catch a certain distinctive smell or taste which returns you to an earlier time which you associate with it. There's a famous example of this close to the beginning of Marcel Proust's famous sequence of novels *A la recherche du temps perdu*. The narrator (Proust himself, thinly disguised) describes how eating cake crumbs mixed with tea makes his childhood incredibly real to him again.

At first he feels an overpowering sense of joy, a sense that 'the vicissitudes of life had become indifferent to me, its disasters innocuous, its brevity illusory ... I had ceased now to feel mediocre, contingent, mortal.'[26] He begins to realise that the source of this joy is that he has transcended linear time and reconnected with the past, back to when his aunt Leonie used to give him a piece of madeleine cake after dipping it in her tea. Immediately he becomes aware of the 'vast structure of recollection' held by taste and smell, as the long dormant past – of his aunt's house, the village square, the streets, gardens and the park – springs to life.

Or this might happen when you revisit a place where you used to live. I experienced this a few years ago when I went back to Coventry, the town I lived in while at university. This is how I described the experience in my diary at the time.

> I could feel everything I felt then so strongly it was like going back in time as well as in place. It was so real that it scared me a little. I stood outside the house I used to live in on Beech Tree Road and I could see myself in it, half expected the door to open and to see myself walk out of it. I felt like a ghost from the future eavesdropping on the life of the person I used to be. The sensation was so strong that it's made me feel that in a way the past still exists now.

Of course, this experience or Proust's *could* be nothing more than especially powerful memories. Toynbee's visions of historical events could just be unusually vivid daydreams or hallucinations. There's no way to authenticate these experiences, other than the subjective feeling of their reality. But again, in terms of the other ideas we've looked at in this book, and in terms modern physics, it would be very surprising if such experiences didn't occur sometimes. If the whole of time is a panorama, and our linear sense of time is only created by our egoic consciousness, it makes sense that at times the barrier should fall away, and enable us to glimpse the reality of the past. These instances of retrocognition suggest that all of the past experiences of our lives – and indeed, all of the events that have ever taken place in the history of the world – haven't just passed away and vanished into dust. In some sense, they all still exist, in parallel with the present. It may not be true that – in Benjamin Franklin's phrase – 'lost time is never found again'. Rather, it may be that, as D.H. Lawrence put it, 'perfect, bright experience never falls to nothingness'.[27]

Precognition and retrocognition in hypnosis

We've already seen – in Chapter 4 – that under hypnosis time can be massively expanded, so that people can accomplish tasks that normally take an hour or longer in minutes. But hypnosis can radically change our relationship to time in another way. It sometimes takes us beyond the bounds of linear time, so that we can – apparently – re-experience the past, and *pre*-experience the future.

Once I saw a stage hypnotist who told one of his subjects that he was five years old again. Immediately the person's voice and language switched to that of a five-year-old. And not only this, he seemed to be living in the same world that he'd lived in as a five-year-old, arguing with his brother, crying because he wouldn't give him any sweets, and complaining to his mother. It was almost as if he wasn't just remembering what it was like to be five years old, but actually *being* five years old again. Beginning with Freud, psychoanalysts have always used hypnotic regression in a similar way, to enable patients to relive painful experiences, and – hopefully – to resolve them.

Regressing a person to previous stages of their own lives sounds impressive enough, but some hypnotists believe it's possible to go even further than this, and regress a person to their previous *lives*. This is highly controversial, of course – in order to accept it you not only have to believe that linear time is an illusion, but also to believe in reincarnation. I've always been sceptical about reincarnation, but I admit that there are some very striking cases of *apparent* past-life regression. For example, there's a well-attested case of a patient who was having hypnotherapy for claustrophobia and fear of water and who, when hypnotised, 'became' a man called James Johnston, who died on a submarine in the Second World War. The patient told the hypnotherapist, Rick Brown – who later described the case in the *Journal of Regression Therapy* – a massive amount of detail about his past life, including the names of his submarine and the crew members, where it was stationed, how and when it sank (22 February 1942), as well as about his life before the war. Brown examined US Navy and Civilian Conservation Corps documents, high-school and birth records, and travelled to James' hometown to interview friends of relatives – and his research found that all of the information his patient gave was correct.[28]

Even more strikingly, an American woman (identified as L.D.) had a series of 36 taped regression sessions with the hypnotherapist Linda Tarazi. In the sessions, L.D. spoke in Spanish (a language she had never learned or spoken before) and 'became' a woman called Antonia who lived in Spain, Germany, England and Peru during the 16th century. The sessions produced over 1,000 pages of information, which Tarazi and other researchers then spent several years examining and verifying, checking 16th-century Spanish texts and the municipal documents of Antonia's home city, Cuenca in Spain. Some of the details were so obscure and precise that no historian would have known them, let alone a middle-aged American women with little or no knowledge of European history. For example, in one session Antonia talked about a dispute between Inquisitor Ulloa and Viceroy Villar, and mentioned that a man named de Prado supported Ulloa. After a great deal of research, these names were discovered in a 16th-century Spanish text. Antonia also gave the names of some of her friends in Cuenca, and when Tarazi visited Cuenca on a research trip she found eight of these in inquisition records and city archives. In one session, Antonia mentioned that a college was founded in Cuenca in the mid-16th century, and that the staff and students used to meet at her inn. But when the historians working with Tarazi attempted to verify this, they couldn't find any reference to a college; the town archivist knew nothing of it either. However, Tarazi was advised to consult texts stored at Loyala University, and there she found an obscure old seven-volume Spanish work which referred to the founding of the college.

Other information which was verified included Spanish shipping laws in the Indies, the types of ships used in the Mediterranean and the Atlantic, the date of the Edict of Faith on the island of Hispaniola, and the names of priests executed in England in 1581 and 1582, and the method of execution. Many of these facts didn't appear to have been published in English at all, and some could only be found in the archives at Cuenca.[29]

And if hypnosis can enable us to re-experience the past, perhaps it can give us access to the future as well. In 1964 the historian Charles Hapgood saw a demonstration in which a psychologist called Kenneth Lyons regressed a hypnotised man to his first day at school. As with the demonstration I saw, the man's voice changed to a child's and he described all the day's events in great detail. However, after this Lyons appeared to regress the man to a previous life. He 'became' a man from the 17th century called Charles, who lived in Cornwall and died in prison in London. Hapgood was startled and impressed, and persuaded Lyons to do tests under more scientific conditions, which led to two years of experiments with past life regressions.

However, at a certain point it occurred to Hapgood that hypnosis might also be used to 'progress' to the future. He began to test this, and obtained some impressive results. For example, a student called Jay was hypnotised and 'progressed' to next Wednesday. He gave a detailed account of everything he was going to do that day and when he emerged from his hypnotic trance he had no recollection of what he'd said. And when, on Wednesday evening, Hapgood asked him about his day, all of the details matched his previous week's description perfectly, right down to the menu in the college canteen, his tests and assignments in class, a chance meeting with a pilot who explained the details of a plane crash to him, and so on. Hapgood repeated the experiment with other students, with the same successful results.[30]

If precognitive experiences often occur in the non-egoic realm of sleep, it's not surprising that these glimpses of the future and the past can occur in the non-egoic realm of hypnosis. However, I should point out that by no means *all* non-egoic states seem to lend themselves to non-linear time experiences. For example, we've seen no evidence that they occur in non-egoic states caused by accidents, emergencies, drugs and schizophrenia, even though time can slow down massively in these situations. In a sense, though, this *does* happen in HSCs, as we saw in the last chapter. At the highest levels of consciousness, when the ego has been completely transcended, we also transcend the linear-ness of time and experience an 'eternal now'.

Near-death experiences (NDEs)

Another non-egoic state in which we may go beyond the bounds of linear time is in near-death experiences. As most people know, these experiences take place when people die for a short time before coming back to life. Their consciousness appears to detach itself from their bodies, and they often find themselves looking down on themselves from above. They frequently see a brilliant white light, and experience a powerful sense of peace and euphoria. Some scientists and sceptics have attempted to account for the experiences in terms of brain activity, suggesting that they are caused by a lack of oxygen in the dying brain, or a rush of endorphins in response to pain. I'm not going to go into all of the arguments against these theories here,[31] but there's one essential point which works against them: in many cases, near-death experiences occur when a person's brain and body are clinically dead, showing no signs of activity whatsoever. And if a person's brain is clinically dead, then it's impossible for them to produce *any* experience, in the same way that it's impossible for a turned-off television to produce a picture.

The question of duration hasn't been given much attention in NDEs – understandably, since the other aspects of the experience are much more striking and significant. But it seems that, in general, people experience a sense of expanded time, or of timelessness. Near-death experiences have many parallels with HSCs and sometimes people have the same sense of transcending space and time which mystics like Meister Eckhart and Richard Jefferies described. For example, a man who 'died' for a short time after crashing his car into a wall, described 'a sensation of serenity, timelessness, [of being] warm and comfortable in complete darkness, floating high above a distant cacophony of indistinct voices'.[32] And here a retired air-force colonel describes how he became aware that linear time is an illusion after he was electrocuted.

> In 1957 when I was 13 years old, I was electrocuted with 7,200 volts and had the classic NDE experience ... The conventional distortion is the construct of a 'past', a 'now', and a 'future' ... At the earliest level of my NDE journey that human construct fell away, that 'arrow of time' ... Then followed a series of increasing levels of consciousness. I clearly recall the joy of additional dimensionality, of wanting to pause and celebrate the arrival of each additional dimension.[33]

Other people experience an extreme 'stretching out' of time when they leave their bodies. They may only be outside them for a few seconds and yet undergo an incredible range of experience – the journey through the tunnel towards a light, encounters with deceased relatives or with beings of light, visions of other realms of reality, and so on. As one of Britain's leading researchers into NDEs, Peter Fenwick, puts it: 'Time is often changed in near-death experiences, and some people describe the period of the experience as being almost an eternity.'[34] For example, in 1986 a woman called Avon Pailthorpe lost control of her car on a motorway. It went into a spin on the fast lane and she was struggling to control it when suddenly she found herself out of her body, moving through a tunnel at high speed. The tunnel was completely dark to begin with, but gradually became lighter and at the same time less defined. As she was moving through it she became aware that a debate was taking place, that presences of some form were discussing whether she should return to her body or not. She felt that there were many minds on both sides, throwing arguments back and forth, and after a while she had the sense that they had made a decision, even though she didn't know what it was. Then, all of a sudden, she was back in her body and back in the car in the fast lane. She had the sense that she 'didn't have permission to go yet, because my children only have one parent'. It felt strange to be back in her body; she felt as though she'd been away for so long that she needed to reacquaint herself with it. And at that moment she saw a white car coming towards her at a high speed and lost consciousness.[35]

The car hit her and caused severe injuries; she was 'deeply unconscious', and at the hospital doctors were afraid she wasn't going to survive. However, one of the strange things about her experience is that it occurred *before* the point of collision. Somehow the normal linear flow seems to have been disrupted, so that she had the experience a few seconds before she actually 'died'. The effect came before the cause, in a similar way to how explosions inside the nuclei of atoms sometimes occur before their cause. But this experience is also interesting in that Avon Pailthorpe could only have been out of her body for a few seconds in clock time – the gap between her car spinning out of control and being hit by an oncoming car – and yet it's clear that, psychologically, she experienced far more time than this.

The most striking aspect of NDEs in terms of time is the 'life-review', in which the person sees all the events of his or her life replayed in a flash. Some researchers have found that between a third and a quarter of NDEs feature a life-review. Here a person I interviewed describes the one he experienced when he died for a short time after sitting on a worn-down power cable at a music festival.

The whole of my life up to that point whizzed by in a series of scenes, everything that I'd ever done and that had ever happened to me. It was all right there in front of me, in amazing detail, including dozens of events that I thought I'd forgotten. I saw so much even though it could only have lasted for a few seconds.

In the following example, the person's sense of time was so extended that, even though the review could only have lasted a few seconds in 'real' time, he not only saw all the different events of his life, but had time to deliberate over each of them. He was an ex-RAF pilot called Allan Pring, who 'died' during a routine operation which went wrong.

Without any anxiety or distress I knew that I was dead, or rather than I had gone through the process of dying and was now in a different state of reality … Although I no longer considered that I had a physical body, nevertheless I felt as if I were floating in a vast empty space. Then I experienced the review of my life which extended from early childhood and included many occurrences which I had completely forgotten. My life passed before me in a momentary flash but it was entire, even my thoughts were included … There was one most peculiar feature of this life review and it is very difficult to describe, let alone explain. Although it took but a

moment to complete, literally a flash, there was still time
to stop and wonder over separate incidents.[36]

Sometimes the life-review occurs in the moments before
dying, or at least when it seems inevitable that a person is
going to die. A Swiss geologist called Albert Heim, for example, fell from a precipice while mountain climbing, and his
'whole life flashed before his eyes' while he was falling.

During the fall a flood of impressions swept over me.
What I felt and thought in those five to ten seconds cannot be told in ten times as many minutes. I watched the
news of my death reach my loved ones and in my
thoughts, I consoled them. Then, I saw as if on a distant
stage, my whole past life playing itself out in numerous
scenes.[37]

Interestingly, there have been cases of people seeing a
review of their *future* as well as their past. They can experience 'flash-forwards' as well as flashbacks. The person sees
the whole of their life as a panorama, the whole of their
future as well as their past spread out in front of them. They
realise that *if* they return to their body that this is the pattern
their future will follow.[38]

These experiences can be seen as examples of the fifth
law of psychological time ('Time often passes slowly, or
stops altogether, in situations where the "conscious mind"
or normal ego is in abeyance'). The important point is that
near-death experiences appear to be experiences of egoless-
ness, and provide further evidence that our normal sense of
linear time is an illusion. Again, the person retains a sense
of identity, a conscious self which is aware of what's happening, but this isn't the normal ego-self; it's the pure
awareness that remains when the ego dissolves. When we

'die' we transcend the ego; we leave it behind along with our body, and as a result it no longer controls our sense of time. The normal rules of sequence and duration no longer apply. And in this realm it's perfectly possible for time to slow down radically or disappear, for decades of a person's life to be replayed in an instant, and for the division between the future, present and past to break down. In this state of ego-transcendence, time no longer exists.

Time is a creation of the ego

It's clear that the timeless realm which trans-egoic states give us access to has many different aspects. Sometimes time might just seem to expand massively, or to stop altogether – as in sportspeople's Zone experiences, or in psychedelic or schizophrenic experiences. At other times, we might transcend linear time and glimpse events from the future or the past – as in instances of precognition and retrocognition, or in a hypnotic trance. More powerfully still, we might see the whole of our own lives as one present moment, as in near-death experiences. While most powerfully of all, we might experience all of the past which has ever been and all the future which will ever be in one present moment – as in the mystics' experiences of an eternal now.

The only way to make sense of these experiences is to think of psychological time as a creation of the ego, which doesn't exist outside the ego in the same way that universal time doesn't exist outside gravitational fields. The reality seems to be that linear time is an illusion, and that the *whole* of time, including everything which has ever happened and everything which is ever going to happen, exists as one unbroken whole, alongside and in the present. As the 18th-century German philosopher Kant suggested, there's no such thing as 'objective time'. Time is only a 'category' of our minds which we need to order our experience. It doesn't exist in the world, we 'impose' it on the world, as a part of our 'apparatus of perception'.[39] Or in the psychologist Kurt Dressler's words: 'In its true essence reality appears to know only unbroken wholeness and all-presence. To our consciousness it shows itself in temporal succession.'[40] And just in case this concept seems a little vague and unscientific, it's important to remember that, as we've seen, it corresponds exactly with some of the most fundamental ideas of physics about the essential reality of the universe.

It's important to look at all the evidence for these ideas in totality. If a sceptic looked at precognitive or NDEs separately, they might just be able to convince themselves that they can be explained in a rational scientific way. But if they were forced to consider the large number of scientific experiments validating precognition; the massive anecdotal evidence for precognition, premonitions and retrocognition; the huge number of strikingly similar cross-cultural reports of NDEs and HSCs; Zone experiences and other non-egoic states under the influence of hypnosis or drugs – and not forgetting the insights of modern physics – they would surely have to accept that our normal straightforward linear view of time is an illusion.

After noting that the linear view of time has no place in modern physics, Paul Davies ponders over whether this is the fault of physics, 'which pays scant attention to the role of the conscious mind in the universe', or whether the truth of the matter is that the passage of time is just an illusion created by our minds.[41] But the evidence we've looked at makes it fairly clear that the 'fault' lies with our minds rather than with physics. They are the only places where time passes.

The important point may be that we *need* to have a sense of linear time to organise our experience and make survival possible. The French philosopher Henri Bergson recognised this. According to him, all of the future and all of the past exist now – but if we had access to this level of information, life would be impossible. We'd be bombarded with so much information that we'd be unable to concentrate on the present. And because of this, says Bergson, our consciousness has developed the ability to 'screen out' the past and the future, and focus our attention exclusively on the present.[42] In other words, our sense of linear time is a survival aid, like the desensitising mechanism we looked at in Chapter 2. They both have the function of 'switching off' our attention to information to help us concentrate on the practical business of survival.

But does this mean that it's inevitable that we're subject to time, since we need a sense of linear time in order to live in the world and organise our lives properly? After all, even mystics have to step back into linear time sometimes, when it's time to cook a meal or to answer the phone. But here it's important to take the example of traditional indigenous peoples again. They seemed to have a more expanded sense of time than us, and a much more pronounced ability to live in the present, while still managing to live in the world, dealing with the practicalities of everyday life. And I believe that it's possible for us to be less subject to time in the same way. It may be possible for us to change the way we perceive time, so that we can slow it down and even transcend it. And we're going to investigate ways of doing this in the last two chapters of this book.

8

Controlling and Expanding Time

Now it's time to turn away from the philosophical to the practical, and look at how we can apply everything we've learned about time to our own lives. In the next two chapters we're going to investigate what we can to control our sense of time passing – in particular, what we can do to slow it down, so that we can effectively increase the amount of time that we're alive. And also, on a level beyond this, we're going to look at what we can do to free ourselves from time altogether.

The keys to this are the five laws of psychological time and what we identified as their causes – the two 'relativities' of psychological time: information, and our ego. Once you know why something happens, it becomes possible to control it. If you know why a certain disease is caused, for example, it becomes possible to find a cure. And in the same way, if we know why time passes at different speeds in our lives, it should be possible for us to control it.

Making time pass quickly

If you want to make time pass quickly, there are three things you need to be aware of. The most important strategy is the one which we all use already: putting ourselves into states of absorption. Secondly, it's important to try to – where possible – avoid sources of pain or discomfort, since they stretch time. And thirdly, it's important to try to make yourself *less conscious* of time, since the more aware of it we are, the slower it seems to pass.

Take the example of a long-haul flight. The most effective way of making the time pass quickly is to get into states of absorption by reading, watching films, chatting, etc. But it's also essential to make sure that you're mentally and physically comfortable. If you're anxious or nervous and feel cramped in your seat and your clothes are too tight or sweaty, you'll find it very difficult to distract yourself from the moment-to-moment reality of the situation. (You could also make yourself less conscious of time by not wearing a watch.) My most comfortable long-haul flight, which also passed the quickest, was one when I managed to have a good meditation before, during and after take off, putting myself into a very relaxed state. I also did some yoga in the waiting lounge, so that I felt stretched and loose in my body. In addition, the plane was quite empty so that I could stretch out, and there were no clocks visible, so I had no idea how much time had passed.

Let's also consider someone who has a job she doesn't particularly enjoy, and so wants her days to pass as quickly as possible. Her main aim should be to make sure her mind is always occupied, that she always has enough tasks to maintain a steady state of absorption. It's fine for her to have short periods of inattention, of course – in fact, she needs these to relax – but if these carry on for too long, she'll slip into a state of boredom and time will start to stretch out.

However, the problem here could be that her job is so undemanding and repetitious that it's very difficult for her to maintain a state of absorption. If this is the case, she might have to make sure she distracts herself in other ways, by listening to the radio or chatting. She also needs to make sure she feels mentally and physically comfortable (which could mean that she's not tired or anxious, or that she has a harmonious relationship with her colleagues), and that she's as unaware of time as possible by not wearing a watch or taking clocks down from the wall.

Longer periods of time

It's fairly easy to make short periods of time pass quickly – say, 10 minutes at the bus stop, an hour in a waiting room or even a day at work. But what if you had a whole year – or even ten years – of unwanted time in front of you? How would you go about killing all of that time?

This is the situation that prisoners find themselves in. As one prisoner told the researcher Diana Medlicott, 'Time passes extremely slowly. I've been in twelve days and it seems like four weeks. Time just doesn't go.'[1] Studies have shown that most long-term prisoners develop strategies of dealing with the unwanted time in front of them. Some give themselves long-term projects, such as body-building, or 'mind-building' through doing courses and getting qualifications. As Stanley Cohen and Laurie Taylor note in their book *Psychological Survival: The Experience of Long Term Imprisonment*, the purpose of this is to impose some sort of order on the frightening empty space of time in front of them, creating stages and 'building their own subjective clock in order to protect themselves from the terror of the "misty abyss"'.[2] In addition, prisoners with long sentences often break the time down and never think further ahead than the next 'chunk' of time. For example, if they have to serve twenty years, they 'do it five years at a time'.[3]

However, as with any short period of time, the most effective thing that prisoners can do to make their sentences pass quickly is to spend as much time as they can in states of absorption, either by working, watching TV, taking up hobbies and interests, and generally taking any opportunity they can to keep busy. In Solzhenitsyn's novel *One Day in the Life of Ivan Denisovich*, based on the years he spent in one of Stalin's gulags, the main character describes 'how time flew when you were working ... The days rolled by in the camp – they were over before you could say "knife"'.[4] But since most prisoners' jobs are likely to be mundane and undemanding, this might not be so easy. Commenting on their study of the inmates at Parkhurst Prison in England, for example, Cohen and Taylor write that: 'It is not much use hoping that a man in Parkhurst's "tag shop" will become involved in his job of sticking metal ends in to the lengths of green string used to keep files together.'[5] However, some prisoners did tell Taylor and Cohen that they felt that working was better than doing nothing, and did make time go faster if they made an effort to immerse themselves in it.

The opportunities to get into states of absorption in prison might be limited in other ways too. Some prisoners are locked in their cells almost the whole day, and don't have any means of distraction apart from television and books. In this situation it's not surprising that, as we noted in Chapter 3, many prisoners try to 'sleep away' their sentences. As a last resort, sleep is the most effective way of all of making time disappear. In a more minor way, however, it might help prisoners a little to remove time-markers like clocks and calendars from their cells, and to try to keep as physically comfortable as possible (e.g. by exercising).

The waiting experience

Waiting is one of the biggest banes of our lives. We all hate it, but unfortunately we have to spend a large amount of our time doing it: at post offices, banks, train stations, airports, dental surgeries, supermarkets, and so on. A 1988 study found that the average American citizen spends around 15 per cent of their life waiting in queues, and about 3 per cent trying to make return phone calls. And as well as causing a great deal of psychological unrest, there are studies showing that this continual waiting is responsible for physical problems too, such as ulcers and coronary heart disease.[6]

Most organisations make a token effort to relieve the burden of waiting, by offering magazines to read or providing TVs or radios. However, the relationship between anxiety and slow-passing time means that it's especially important for waiting conditions to be as pleasant and relaxing as possible. In a recent paper, Dr Stephen Buetow of the University of Auckland looks at the factors which affect time perception, and suggests ways of accelerating time for waiting patients. He emphasises the importance of creating a comfortable environment and employing staff with a friendly and empathic manner.

> Practices can compress time by establishing physical and social environments that, as far as possible, make waiting less of an imposition … To create this environment,

> staff who interact with patients need to demonstrate a
> warm, caring manner, not merely avoid harm …
>
> The physical environment of the waiting room –
> including materials, colours, shapes, textures and
> sounds – should be familiar and reassuring to the partic-
> ular patient clientele, though not monotonous.[7]

However, there's another method of dealing with frustrat-
ing waiting experiences: trying to live *in the moment*. When
you're waiting, part of your attention is focused in the
future, on getting where you want to be – being treated by
the doctor or dentist, packing your bags at the checkout, or
sitting on the bus or the train. Apart from this, your mind is
full of the usual chattering thoughts about what you did last
night, what you've got to do later, etc. – the kind of information which, as we've seen, is responsible for slowing down
our perception of time. But instead of focusing on the
future, or on your thought-chatter, you should concentrate
your whole attention on the *now*. In a waiting room, you
should be aware of the furniture and other objects around
you, take in the is-ness of everything you can see, the
colours and shapes and textures, the light coming through
the window, and the trees blowing in the wind outside. Be
aware of your own presence in the room, the feeling of your
body against the chair, and your breathing. Or in the queue
at the supermarket: pay attention to the mass of different
colours and shiny packages on the shelves, and all the different people unloading their trolleys – and again, be aware
of your own presence there. This may not lessen your perception of the time you spend waiting – in fact, it will have
the opposite effect, and slow down time. But living 'in the
now' like this is such a positive and pleasant experience that
it doesn't matter how much time we do it for. In fact, as

we'll see in the next chapter, in a way, living in the now means going beyond time altogether.

Just to summarise this section, then, if you want to make time pass faster in any situation, you should:

- spend as much time as you can in states of absorption, e.g. working, watching television, reading, doing 'flow' activities such as gardening or painting
- minimise physical and psychological discomfort, making sure that you're not in an anxious or frustrated state
- take off your watch and avoid looking at clocks, so that you aren't aware of time passing.

Making time pass more slowly

Holidays, weekends away, an evening without the kids, a few days with your lover before one of you has to go away, a prisoner on day release, a soldier on leave – these are all situations in which we might want time to pass as slowly as possible. And the keys to doing this are the second and fourth laws of psychological time: 'Time slows down when we are exposed to new experiences and environments' and 'Time passes slowly in states of non-absorption'.

If you don't want time to pass quickly, you should try to avoid getting into states of absorption. Sometimes when I have an evening on my own to relax, when we've put our children to bed, my wife has gone out and there's no work to do, I purposely decide not to watch any TV programmes or DVDs or to read magazines because I want the evening to last for as long as possible. I don't want to put a film in the DVD player or open up my new copy of *Mojo* magazine and emerge out of a timehole two hours later, when it's almost time for bed, where there's another timehole waiting for me before I have to go to work the next morning. I want to savour my free time, make it last for as long as possible. So instead I usually listen to music for a while, spend some time reading a book (usually a non-fiction one), phone a friend for a chat, or have a stroll around the garden. It's true that I'll still probably spend *some* time in a state of absorption – especially when reading – but this will be nothing like the intense absorption I'd experience if I spent the whole evening watching TV. And in this way I experience more time in the evening. By the time I go to bed I feel like I've had a long and full evening, packed with different experiences. I often do this on Sunday evenings as well. I want to make the most of my last evening off before going back to work, to make it last for as long as possible, and so I avoid watching television or reading magazines.

The same holds true for any period of time – if you want a weekend away or a week's holiday to last for as long as possible, then spend as little time as possible in states of absorption. If you're one of those people who go away on holiday and spend their days on the beach with their nose inside a best-selling novel, or who still watch their favourite TV programmes in their hotel room in the evenings, then you shouldn't really complain if the days seem to go quickly. You could make your holiday last a lot longer by limiting the amount of time you spend reading on the beach or watching TV, and spending more time walking through the countryside, exploring different towns and the streets of the town where you're staying, and chatting to people.

Of course, the second law of time is important here too. Time usually does go slower than normal when we go on holiday, because of the unfamiliarity of the environment. However, many of us limit the effect of this by going to places which aren't *particularly* unfamiliar to us, which incorporate many familiar aspects of our home life. We holiday in tourist towns full of other English people, where most of the locals speak English, where restaurants serve English food and bars serve English beer. This familiarity makes us feel comfortable, but it's also the reason why we usually feel that, after maybe a week of moving slowly, time starts to speed up again, and the last few days of the holiday seem to flash by. There's such a limited amount of unfamiliarity in these environments that it only takes a few days for our minds to become desensitised to it. If you really want to have a holiday that lasts for a long time you need to go to a completely unfamiliar environment, a place where people don't make any concessions to your culture, where they don't speak your language or cook your food and where the architecture and vegetation is completely different. Go on an adventure holiday instead of a tourist one. Go on a walking tour through the Spanish countryside instead of the Costa del Sol. Go to Asia instead of Europe.

You could also use the second law to expand time for a weekend or a day. If you're going away for the weekend and want to make it last for as long as possible, make sure you go to a place you've never been to before, or do something completely new to you. My wife and I became aware of the benefits of this when she was off on maternity leave. At the time I was only teaching three days a week, which meant we had a block of four days off together. After a while we noticed that if we spent those four days at home they passed quite quickly, but if we were more adventurous and had at least a couple of days out – visiting relatives or friends in different towns or going walking in the countryside or by the sea – we realised that time slowed down. When I went back to work on Monday we always felt that it had been an extremely long time since I left the previous week.

However, in most situations it's best to think in terms of using the second and fourth laws together. If two lovers spent their last few days together at home watching TV and playing computer games, the week would pass incredibly quickly. Instead, they could try to expand time by unplugging the television and the computer and cancelling the morning paper, trying to spend as little time as possible in states of distraction. In addition, they could expand time by getting away from their home environment, going abroad or even just having a few days out to expose themselves to unfamiliarity.

There's also, however, a third way of slowing down time, which we touched on at the end of the last section: by living in the now. But since this is the main theme of the next (and last) chapter of this book, we won't investigate it further at this point.

Slowing down time over our whole lives

Hopefully this book has made it clear that it's not sufficient to just measure time in terms of hours, months or years. You also need to include a person's *experience* of time; their subjective perception of it. In a sense the normal way in which we judge the length of a person's life is misleading. How much time we experience in our lives doesn't just depend on their length in calendar time. In my opinion, it mainly rests on our perception of time passing, which in turn depends on how we live our lives and the state of consciousness in which we live them. In terms of psychological time, it's quite possible for a person who dies at the age of 30 to live through much more time than a person who dies at the age of 80.

In fact, because of the way we live our lives, I believe that most of us experience far less time in our lives than we potentially could. Of the five laws of psychological time, we're continually subject to the first ('Time speeds up as we get older') and the third ('Time passes quickly in states of absorption'). We rarely experience the second law (on new experiences) and fifth (on ego abeyance). In other words, the laws which make time pass faster affect us the most often, and the ones that make it pass slower only very rarely. (It's true that we're often subject to the fourth law – on non-absorption – but we try to experience this as little as possible.) Most of us spend our lives in familiar environments, living through familiar experiences. This means that the amount of perceptual information we process gradually decreases through our lives, due to the desensitising mechanism. We also spend as much as we can in time-constricting states of absorption, because we find states of *non*-absorption so unpleasant.

It's quite easy to apply the laws of psychological time and the laws behind them to our whole lives and to expand the time that we experience. This is an alternative (or complement) to trying to extend our lives in terms of years by eating healthy foods and exercising.

In theory, you could live through a lot more time in your life by trying to spend as little time as possible in states of absorption and distraction. You could even make a conscious effort to do things which you find boring, in spite of the psychological discomfort this brings. In his book *Space, Time and Medicine*, Larry Dossey tells the story of a doctor colleague of his who noticed that many of his patients took up fishing after they were diagnosed with cancer. He believed that this was all about expanding their perception of time: 'If you're sitting on a boat doing nothing but waiting for a fish to bite, time drags. I can't think of any better way of making the days longer. It's a perfect recreation for someone who believes he's going to die and that his time is limited.'[8] However, the doctor's reasoning may be slightly faulty here – fishing can have a time-stretching effect, but not so much because it's a boring activity, as he implies, but because it concentrates your attention fully on the present moment. But the principle is still true: if you want to experience more time, you could swap your stimulating job for a boring one, give up your car so that you have to spend a lot of time waiting for buses or trains, give away your TV and your books so that you have to spend your evenings pacing around your flat with nothing to focus your attention on.

In reality, none of this would work. There's obviously no point living for as long as possible if your life is miserable, and in any case, the psychological discomfort of boredom would never let us endure these situations for long. It's difficult to minimise the amount of time we spend in states of absorption without actually changing *ourselves*, since we need to keep our attention absorbed to escape mental discontent and boredom.

However, this is a question of degrees, and it's probably still possible to extend our lives to a *limited* extent by avoiding absorption. This means making a conscious effort *not* to kill time by working more than you need to, watching television, playing computer games and so on. It means getting used to *not* having any external things around you to put your attention into – getting used to just living in the moment, and giving your attention to your surroundings and to what you're doing. Or to a lesser degree it might just mean choosing milder kinds of distraction which don't put you into a state of complete suspended animation and don't make time disappear to the same extent. You could read books instead of watching TV, for example, or play chess or football instead of your Gameboy or PlayStation.

However, there's a much more effective way of changing our lives so that we experience more time in them – and it comes through making better use of the second law of psychological time ('Time slows down when we are exposed to new experiences and environments').

A short life within a long lifespan

Let's take the example of a fairly 'normal' person who dies the age of 75. Let's say he spent all of those 75 years in the same town, leaving it only for occasional holidays. He left school at the age of fourteen, did an apprenticeship as a plumber, and spent the whole of his working life doing that job, perhaps working for two or three different companies. He got married at the age of 25 – to a girl who lived close by – and had two children. They lived in a small terraced house for their first ten years of marriage, then moved to a larger semi-detached one a mile or so away, where they stayed permanently. During his younger years, the man worked long hours and didn't have much free time, but when he did have some he liked to listen to the radio, play with his children or – at weekends – go drinking with his friends at the pub. Later, when his children left home, he had more time on his hands and spent most of it watching television, doing jobs around the house and pottering around in the garden. His life mainly consisted of working, sleeping and watching TV until he retired at the age of 65. After that he filled his days with more gardening, more television, more jobs around the house, as well as helping his wife with the housework, until he died ten years later.

This is, of course, exactly the kind of life in which time passes frighteningly quickly. By the time he was in his fifties, the man probably felt that the years were going by more quickly than he could count. He spent his whole life in the same environment, in the same house in the same area. And his life consisted of a repetition of the same experiences – the same daily, weekly and yearly routines repeated again and again. He very rarely exposed himself to any unfamiliar environments or any new experiences, and as a result his consciousness only absorbed a trickle of perceptual information, a trickle which grew smaller and smaller each year. As well as this, he spent most of his life in states of absorption – states of passive absorption when he listened to the radio and watched TV, and states of active absorption when he worked at his job, did jobs around the house and other practical tasks. This stole a great deal of time from his life too.

In other words, throughout his life, the man very rarely experienced the second law of psychological time (new experience) but was constantly affected by the first (time speeds up as we get older) and the third (absorption) laws. As a result, it's possible to say that although he was on the surface of the earth for 75 of its orbits around the sun, in terms of psychological time his life was actually quite short.

A long life within a short lifespan

We can compare this person's life with a friend of my wife's called Andrea, who died a few years ago at the age of 31.

My wife grew up with Andrea. They lived on the same road and were best friends all the way through school. But from the age of eighteen to 28, Pam (my wife) hardly saw her, because she was constantly travelling. She left home to go to university, but dropped out after a year and went travelling with a boyfriend, planning to stay away for as long as possible. They went to Greece, where Andrea worked in a bar, and ended up getting jobs on a luxury yacht, where she worked as a belly dancer/waitress. She split up with her boyfriend after a while but carried on travelling. She was incredibly self-confident and had a gift for instantly making friends, and was always going on new adventures with new people. She lived in an ashram in India for two years, before moving on to Hong Kong, where she worked in a bar again. From there she went to Japan, where she did some English teaching, then back to India for a few months' travelling, then on to Thailand ...

By 1996 Andrea was living in Spain, on a spiritual commune. Things didn't work out there though, or perhaps she was finally getting fed up with the rootlessness and insecurity of her lifestyle. In 1997 she came back to England and told her family and friends that she planned to settle down now. She decided to study again, and after a few months doing temporary jobs and living with friends, she went back to university. Amazingly, this time she managed to keep her restlessness at bay and finish the three years of the course.

But by that time she'd decided that settling down wasn't really for her after all. She did a short TEFL (teaching English as a foreign language) course, applied for jobs abroad, and was offered one in Korea. Everything was set – she booked a flight and had a leaving party the Saturday before she was due to go. But the day before the flight she suddenly lost consciousness and collapsed while cycling. She had a brain haemorrhage, and was already dead when the ambulance arrived.

This is incredibly tragic, of course, but in some ways Andrea's life and death was *less* tragic than that of the old man we've just met, because it's likely that she lived through *more time* than him, even though her life was much shorter in terms of years.

Her life was full of newness and unfamiliarity. She kept exposing herself to new environments and experiences. She was so restless that as soon as she started to get used to one situation or environment, she had to release herself from it and jump into a new one. In this way she never gave the desensitising mechanism a chance to take hold, and to reduce the amount of perceptual information she processed. During her ten years of travelling, she was probably 'awake' to the raw experience of the world almost constantly, and so time would have passed extremely slowly to her.

Her way of life also probably meant that she spent very little time (at least compared to most people) in states of absorption and distraction. As we've seen, we have to spend time in states of absorption because the alternative is to be bored and to experience the psychic entropy of our thoughts. But when you expose yourself to new experiences and new environments you're *never* bored. You're fascinated (or at least interested) by the newness around you, and so don't need distractions. You live in the moment and in the world, in a 'third' state of being, rather than alternating between states of absorption alternating with states of boredom.

Exposing yourself to unfamiliarity

This, then, is one way you could go about slowing down time: by exposing yourself to as much unfamiliarity as possible. What this really means is seeking out new environments (or experiences) every time your desensitising mechanism starts to change the environment you're *presently* in to the familiar. As the philosopher Jean-Marie Guyau put it: 'If you want to lengthen the perspective of time, then fill it, if you have the chance, with a thousand new things. Go on an exciting journey, rejuvenate yourself by breathing new life into the world around you.'[9]

There are three variations of this kind of life. The first is the way of life that Andrea chose – a life of constant travel, of never settling down at all, of uprooting yourself from one life-situation as soon as the desensitising mechanism starts to take hold and putting yourself into a completely new one, with a new job, a new environment and new people.

One of the most extreme examples of this kind of life was that of the French poet and traveller Arthur Rimbaud, which was all the more remarkable because he lived in a time – the mid-to-late 19th century – when people rarely went more than a few miles beyond the village they were born in. From the age of sixteen, Rimbaud had 'wind in the soles of his shoes', as his poet friend and lover Verlaine wrote, and found it impossible to stay in the same life-situation for more than a few months. He found his hometown Charleville so oppressive that he felt he was 'dying and melting away in the dullness, the drabness and the foulness around me'.[10] As a teenager he ran away several times, but finally escaped for good at the age of nineteen. He went to London and worked as a teacher, then studied German in Stuttgart. From there he walked to Italy, where he worked as a dock labourer. Back in Paris, he enlisted for the Dutch army, and sailed with them to the Sunda Islands. But as soon as the ship set to port he deserted and fled to Sumatra and Java. From there he worked his way back to Cyprus, where he worked as a quarry hand, then travelled to Africa, where he remained for the next few years. He became a trader and gun-runner, and a close friend of the King of Shoa. He was the first European to enter certain parts of Ethiopia and achieved some fame as an explorer. However, when the French Geographical Society contacted him to ask for details of his journeys, he didn't bother replying. He showed the same indifference when he learned that his poems had been published back home and he'd become famous as a 'lost poet'.

Rimbaud died of cancer at the age of 37, but his adult life was so full of new experiences and perceptions that he lived through much more time than most people who live to the age of 80 or 90.

There's a basic problem with this way of life though, which Rimbaud himself encountered. It conflicts with some of our most basic needs as human beings – the need for territory and security, for a partner, for a stable network of friends, and so on. A life of constant travel means that these needs are difficult, if not impossible, to satisfy. It's a little like jumping straight to the top of Abraham Maslow's hierarchy of needs – attempting to satisfy the highest need for 'self-actualisation' without dealing with the lower needs (for safety, love, belonging, esteem, etc.) first. As a result, perpetual travellers may struggle with a sense of dissatisfaction and incompleteness. And in fact this is why most travellers become disillusioned eventually, and long to settle down – although when they *do* settle down, their wanderlust often gets the better of them again.

The second option is a slightly milder alternative: to have a stable life-situation and use it as a base from which to make forays into unfamiliarity. This might mean owning a house in the town you were brought up in, with friends and family and even a partner, and at the same time having a job which gives you access to unfamiliarity. You might be a contract worker, for example, who travels around the world working for a few weeks or months in different countries, with a few weeks at home between each trip; or one who's able to spend half their year working, then the next six months travelling. In this way, you'd still expose yourself to a great deal of unfamiliarity – and so live through a great deal of time – without necessarily sacrificing your basic human needs.

One person who followed this path was the writer Bruce Chatwin. He often identified himself with Rimbaud (he even called one of his books *What am I Doing Here?* after a famous passage from one of Rimbaud's letters) and is often compared to him, but there's a fundamental difference between them in that, whereas Rimbaud was completely rootless in his wanderings, Chatwin always had a home base from which he made his regular forays away. As he said himself: 'No man can wander in fact without a base. You have to have a sort of magic circle to which you belong. It's not necessarily where you were born or where you were brought up. It's somewhere you can identify with, to which you always happen to go back.'[11]

After he left school Chatwin got a job at Sotheby's. This partly satisfied his wanderlust, since it meant occasionally flying to different countries to examine antiques, and he used his long summer break to go travelling through Asia and Africa. But he wasn't the kind of person who could be tied to a job and resigned at 26. He began studying anthropology at Edinburgh University but was too restless to sit out the three years of his course. He made long trips to Afghanistan and West Africa and then took a job as an arts advisor for the *Sunday Times* magazine. But even this was too restrictive for him – after two years he sent the paper a telegram which read: 'Gone to Patagonia for six months.' His trip there led to his book *In Patagonia*, and from then on, he managed to make a living as a writer and could travel completely freely. He thought of himself as a nomad and was on the move so much that his friends were never sure where he was. Even when he was writing his books he could never stay in the same house for long and moved from one friend's cottage or one hotel to another. He always came back to his home in Oxford, and to his wife and friends, but as soon he was there he began plotting his next escape. As his biographer Nicholas Shakespeare writes: 'He could not stay anywhere long before "the malaise of settlement" crept upon him. Even his favourite places soon bored him.'[12]

Chatwin died of AIDS at the age of 48, but like Rimbaud, in terms of psychological time his life was enormously long.

Even this option isn't very practical though, since most of us don't have jobs which allow this kind of freedom. However, it's still possible to expose yourself to a degree of unfamiliarity while living in the same life-situation and environment. This is the third and mildest variation of using this method to expand time in your life. The important thing here isn't the environment you live in, but the experiences you have in that environment. You could expose yourself to unfamiliarity in your working life, for example, by making sure that you change jobs every so often, or at least transfer to different positions within the same company. You could expose yourself to unfamiliarity in your leisure time by always exposing yourself to new areas of interest, knowledge and information, and by learning new skills and encountering new people. And at the same time, of course, you could use your weekends and your holidays to expose yourself to unfamiliar environments. The important thing is to avoid repetition and routine as much as possible, to continually find new experiences instead of just repeating old ones. Living like this won't have the same extreme time-expanding effect of living like Rimbaud, Andrea or Bruce Chatwin, but you'll certainly find that your weeks and months contain more time than before.

One person who lived this kind of life was Jean-Marie Guyau, who I quoted earlier. The French philosopher died of tuberculosis in 1888 at the age of 34, and spent most of his adult years settled in Provence. But although he didn't expose himself to much unfamiliarity through travel, Guyau had an incredible intellectual curiosity. He wrote ten books and numerous articles on a massive range of subjects, including philosophy, psychology, sociology, religion and aesthetics. It's probable that his constant thirst for new knowledge and information, as well as the constant stream of his own new ideas, stretched time for him, and meant that his life was longer than its short span suggests. As the historian of psychology Douwe Draaisma remarks:

> You might say that he lived a longer life by renewing his inner world at every turn. His swift, almost obsessive journey through the most divergent branches of philosophy and psychology must have had as mind-stretching an effect as a genuine voyage.[13]

The problem with unfamiliarity

Just to summarise this section, then, if you want to make time pass slowly, either for a short period or over your whole life, you should:

- avoid states of absorption
- expose yourself to as much unfamiliarity as possible, through new experiences and new environments.

The main problem with attempting to slow down time by exposing yourself to unfamiliarity, however – in any of these three variations – is that you're really just *escaping* the problem rather than actually dealing with it. Like someone who's on the run from the police, this just keeps you one step ahead of the desensitising mechanism instead of addressing the underlying causes. Ultimately this brings a sense of futility too, of being helplessly propelled from place to place with no real reason to be anywhere. The desensitising mechanism is bound to catch up with us at some point anyway, you might say. After all, the world is a finite place, you can't keep travelling to different countries and cultures forever. Eventually the unfamiliarity of strange places will itself become familiar to you, so that you'll be as bored with them as you were with your life-situation at home. In fact, this seems to be exactly what happened to Rimbaud towards the end of his wanderings. 'What am I doing here?' he wrote to his family from Africa.

> What's the use of all this to-ing and fro-ing, these exertions and adventures among strange peoples, and these strange languages one fills one's head with, and these unspeakable ordeals, if one day, a few years from now, I can't rest in a place that's more or less to my liking, and find a family.[14]

But fortunately there's another way of approaching the problem – a more internal approach. We can expand – and even transcend – time through changing *ourselves* rather than just the activities and environments in our lives.

9

Expanding and Transcending Time

If we know that time speeds up as we get older because of our desensitising mechanism, then surely one way of permanently expanding our sense of time would be to stop this mechanism working – or more strictly, to take away the necessity for it to function. Similarly, since the rapid passing of time that we usually experience, and our strongly linear sense of time, are both generated by the ego, another way of expanding our sense of time would be to transcend the ego, to blunt the strong sense of separateness which isolates us from the rest of reality and creates the illusion of time.

As I suggested at the end of Chapter 5, in a sense this would mean existing in a similar state of being to traditional indigenous peoples, with their heightened perception of the phenomenal world and their less pronounced and individualistic egos. And, as I concluded in Chapter 6, this would also be equivalent to living in a permanent HSC, since intensified perception and a transcendence of ego-separateness (bringing a sense of union with the cosmos) are the characteristics of higher states.

The idea of transforming ourselves in this way might seem far-fetched, but in reality it's quite straightforward. At least, the principles are straightforward – actually putting the principles into practice requires quite a lot of self-discipline and effort.

Meditation

The most important single thing we can do to expand and transcend our sense of time is to regularly meditate.

Meditation seems to be natural to human beings – at least, natural in the sense that the practice arose independently in many different parts of the world. Almost every culture over the Eurasian landmass has developed a form of meditation. In ancient India the practice was an integral part of spiritual paths such as Vedanta, Buddhism, Jainism and Yoga, while in China the practice of *tso-wang* – 'sitting with a blank mind' – became popular during the latter part of the first millennium BC. In Islam and Judaism, the spiritual traditions of Sufism and Kabbalah both developed forms of meditation, while in the Eastern Orthodox Church the Jesus Prayer ('Lord Jesus Christ, son of God, have mercy on me, a sinner') is clearly a meditative practice. Some modern Christians are opposed to meditation, believing that quietening the mind leaves it somehow 'open' to the devil, but even in Western Christianity there's a long history of the practice, especially in the monastic traditions. The 14th-century Christian text *The Cloud of Unknowing* (written anonymously by a man who lived in the Midlands of England in the 14th century) recommends a form of mantra meditation, as do the famous spiritual exercises of St Ignatius.

There are probably thousands of different techniques of meditation, but they all follow the same basic principle. The aim is to quieten the mind, to slow down the chattering thoughts which continually whiz through it, until eventually they fade away and we experience a state of mental stillness, or pure consciousness. In meditation we do this by *focusing the mind*, by trying to keep our attention fixed on a certain point, object or image. Different techniques recommend different objects of concentration: a mantra, our breathing, a point inside our own consciousness, a mental image or a real object such as a candle flame or a sacred symbol such as a mandala. Focusing our attention in this way can lead to mental quietness. Our normal thought-chatter is fuelled by attention we give to it, so if we're immersed in it, our minds keep spinning from one association to the next. But if we concentrate on another object, it naturally fades away, like a fire which has run out of fuel. Some practices recommend letting go of the focusing device when the mind has become quiet, so that we can experience mental stillness and peace without any obstruction.

For such a simple practice, meditation is incredibly effective on a variety of levels. Studies have shown that it can bring an almost dizzying array of physical benefits. It can reduce heart rate and blood pressure and has been successfully used to treat hypertension and cardiovascular disease. It can also stimulate the immune system, reduce chronic pain, help alleviate asthma, and even aid recovery from cancer.[1] If any drug had these effects it would quickly become known as a wonder drug and find its way into every bathroom cabinet in the world. But on top of all this, meditation has powerful psychological benefits. Many studies have shown that regular meditation leads to reduced anxiety and stress, and can help people overcome insomnia, depression, alcoholism and drug addiction. More generally, the practice has been found to enhance psychological well-being and increase happiness. In 2003, a study by Paul Ekman of the University of California showed that long-term Buddhist meditators had less activity in the amygdala, the part of the brain linked to fear and anxiety. Ekman found that the meditators were less prone to confusion, shock and anger than other people, suggesting that they were significantly more serene and content.[2] Also in 2003, participants in a study at the University of Wisconsin were given eight-week training courses in mindfulness meditation, which they practised for an hour a day. After the training they had brain scans, which showed that – compared to a control group – there was significantly more activity in the left side of the prefrontal region of their brains, an area associated with lower anxiety and a more positive emotional state. (Interestingly, this study also found that the meditators had stronger immune systems than the control group. At the end of the eight weeks, both groups were injected with a flu vaccine. Tests a few weeks later showed that meditators had responded better to it, producing a significantly larger number of antibodies.)[3] Again, these benefits are probably much greater than Prozac or Valium or any other mood-altering drug, and it's not surprising that many doctors now recommend meditation to their patients.

But most importantly for readers of this book, meditation is – as I mentioned in Chapter 6 – a reliable way of inducing low-level HCSs. A good meditation – in which you sit quietly for 20 minutes or more, keeping your attention fixed to a mantra or your breathing until your mind becomes relatively still and empty – always generates a heightened intensity of perception and weakens your ego structure, resulting in an expanded sense of time. The practice creates an intensification of energy inside you which remains after your meditation is complete. This means that there's no need for the desensitising mechanism to conserve energy by switching off our attention to the reality of the world. Instead, there's energy available for you to be put into perception, so that you're awake to the is-ness of the world around you in the same way that children are. And since the ego is sustained by thought-chatter, when meditation makes our minds quieter, the ego becomes weaker, and may perhaps even fade away altogether.

However, this isn't just about *expanding* our sense of time – it can mean transcending time altogether. If your mind is still and empty during or after meditation, and your attention is wholly focused on the present moment, in a sense linear time ceases to exist for you. There's no past and no future, just a constant state of nowness. I have suggested that our linear sense of time is created by thoughts, because so much of our thought-chatter is oriented around the past and the future. And so when thought comes to a halt, and we're still fully conscious in the present moment, there's effectively no more time. The ego temporarily ceases to exist as a structure, and so does time.

These are the temporary effects of meditation, but meditation works on a long-term basis too. Over time, meditation trains the ego to be quiet. Whereas its normal state is to chatter away wildly of its own volition, filling our minds with a constant stream of thoughts which we don't want to think, regular meditation *tames* the ego, until a state of greater stillness becomes the norm. The mental energy which we normally give away in thought-chatter is conserved, resulting in a permanently high concentration of energy inside us – leaving us potentially awake to the is-ness of the world at all times. We therefore process much more perceptual information than in our normal desensitised state, and so slow down time.

The other long-term effect of regular meditation is that, as our minds become more still and free of thought-chatter, the ego becomes less powerful and defined as a structure. It becomes less dominant in our mental space, and the sense of separateness it creates begins to fade away. We no longer exist in a frightening state of ego-isolation in our heads, in duality to other people and the rest of the world. The walls begin to melt away, creating a sense of communion with the world. And again, since our normal sense of time is generated by the ego, this weaker ego structure brings a permanently expanded sense of time, and a greater ability to transcend linear time by living in the now.

On top of all its other massive benefits, meditation is therefore a powerful tool for transforming our own psyche, for transcending the disharmony of our normal state. But as you'd expect, these benefits don't come easily. Although the technique of meditation is simple, the actual practice can be very difficult. Most teachers recommend two meditation sessions per day, of between 20 minutes and half an hour, preferably sitting cross-legged on the floor in silence. (Some teachers say it's enough just to sit straight up in a chair; lying down isn't advisable though, since there's a tendency to fall asleep.) Some people begin to gain benefits straight away, while others are shocked by the sheer chaos of the thought-chatter they come into contact with and find it almost impossible to slow it down. These people might quickly decide that meditation isn't for them, but it's important to remember that it's a gradual process. It might take months to quieten your thought-chatter and experience any degree of real mental stillness. I've been meditating on an (almost) daily basis for over ten years now, and I'm still amazed at how busy my mind is and how difficult it can be to slow down the whirls of thought. I like to meditate for at least half an hour because it usually takes me at least 10 minutes to disperse the mental fog and feel any sense of inner stillness. (Often though this depends on the circumstances of my life; if I'm in a fairly relaxed state and not too busy or stressed, it can be easier.)

The point is that meditation requires effort and discipline. It's not for dilettantes or for the weak-willed. But if you're prepared to keep trying, the positive effects always come eventually, and the longer your practice continues, the greater they become – until, perhaps after a few years of regular meditation, you might manage to quieten your ego for good, attaining a permanently high level of mental energy and a permanently expanded sense of time.

And it bears repeating that – unlike the mystical perception and the expanded sense of time which come with drug experiences or schizophrenia – the higher state of consciousness which meditation brings *has no bad side effects*. As we've seen, research shows that long-term meditators suffer from fewer health problems, less anxiety and worry, and are happier than other people. They usually have better powers of concentration, are less disturbed by negative emotions like anger and bitterness, and become more caring and compassionate. The important point, again, is that the ego is not destroyed, just *quietened*. It's still there, performing its essential organising and integrating role, but it's no longer as powerful as a structure, no longer as chaotic and as separate.

Mindfulness

After meditation, the second main internal way in which we can expand and transcend time is by practising what Buddhists calls *mindfulness*. We've touched on this a few times already: mindfulness means paying complete attention to the present moment, to your surroundings and your experience, and being fully aware of yourself at that moment and within those surroundings. Meditation creates a state of natural mindfulness, both temporarily and in the long term; our minds become free of thought-chatter, and we spontaneously give complete attention to our surroundings and experience. But it's also possible to be mindful just by making a conscious effort to be, simply by *remembering* to live in the now.

In his book *Coming to Our Senses*, Jon Kabat-Zinn endorses this approach. He discusses the computer scientist Ray Kurzweil's theory that time speeds up as we get older due to the decreasing number of noteworthy events in our lives (which I briefly described in the notes to Chapter 2), and suggests that one way of slowing down time is to 'fill your life with as many novel and hopefully "milestone" experiences as you can'. In other words, you should do what I suggested in the last chapter: expose yourself to as much unfamiliarity as possible. However, Kabat-Zinn suggests that a better method is not to change the way you live, but the way you *perceive* the world. You don't have to chase after adventure to create 'noteworthy' events, you can make all of your ordinary experience noteworthy simply by paying attention to it. As he writes:

> The tiniest moments can become veritable milestones. If you were really present with your moments as they were unfolding, no matter was what happening, you would

discover that each moment is unique and novel and therefore momentous. Your experience of time would slow down. You might even find yourself stepping out of the subjective experience of time passing altogether, as you open to the timeless quality of the present moment.[4]

In other words, being mindful in this way expands time because it means that we're processing much more perceptual information, just as we do in unfamiliar surroundings. We take in all the information that we normally miss because our minds are full of thought-chatter. When we do something unfamiliar or unusual, we *provoke* ourselves into mindfulness. But this shows that we don't actually need the new experience. All our experience can become fresh and new if we make an effort to be mindful.

The best thing about mindfulness is that you can do it at any time. You don't have to put half an hour aside for it, or make sure that you're in a quiet environment. Every daily activity – having a shower, brushing your teeth, making a cup of tea, washing the dishes or even walking to the shops – can be an opportunity for mindfulness. On my courses, I usually suggest choosing two daily activities to carry out mindfully. In the shower each morning, for instance, don't let your mind chatter away about the things you've got to do that day or what you did last night. Instead, try to bring your attention to the here and now, to really be aware of the sensation of the water splashing against and running down your body and the sense of warmth and cleanness you feel.

Or you might choose walking home from the tube station in the evening. Rather than mulling over all the problems you've had to deal with at work or daydreaming about the girl sitting opposite you on the train, focus your attention outside; look at the sky, at the houses and buildings you pass, and be aware of yourself here, walking among them. The most important thing is to stop thinking and start *being aware*, to live in the here and now of your experience instead of the 'there and then' of your thoughts.

I often ask students to do mindfulness exercises chosen from a set of cards. They're divided into different senses, and contain instructions like: 'Sight: Look at your hands, paying attention to the different lines and the different textures and shades'; 'Taste: Eat a sweet carefully and slowly, paying attention to the different flavours and sensations'; 'Touch: Go into the garden outside and carefully feel the trees and plants and the walls'; or 'Smell: Go into the garden outside and carefully and gently smell the trees and plants'. Or sometimes I just bring along a bag of tangerines or sweets, or give everybody a glass of water, and ask them to eat or drink as slowly and carefully as possible. I usually let the students do these activities for around 5 minutes, and they always tell me that they're amazed at how rich and refined their senses become, at how much detail they become aware of and how many different sensations they experience. I don't usually ask them to guess how much time has passed, but when I do most of them overestimate it, suggesting that it's been moving slowly for them.

The students also usually tell me that after the exercises they feel calm and serene, that they have a sense of mental alertness and yet feel relaxed at the same time. This is one of the most striking things about mindfulness: it makes any activity enjoyable. I used to hate washing dishes and did everything I could to avoid it, but once I decided to practice 'mindful' washing up my attitude changed completely. I started to do the dishes in silence, instead of listening to the radio or to music, and instead of thinking I focused my attention on what I could feel and see, the sensation of the hot water against my hands, the splashing of the water and the bubbles and the reflected light on it. I washed each item slowly and carefully, placing them gently on the draining board, instead of rushing through them as quickly as I could. I was amazed to find that now I enjoyed everything about washing up, that it gave me a kind of serene glow inside, and since then I've always seen it not as a chore but as a time for relaxation, a time to slow down and focus on the present.

And as with meditation – and as Jon Kabat-Zinn suggests – this may go beyond just expanding time to 'stepping out' of time altogether. On one level, in moments of mindfulness we transcend linear time because the past and the future cease to be created by thought; nothing exists apart from the present. Time no longer seems to consist of moments which arise and then disappear like an ever-changing river, but of a static present which is always *there*. And on another, deeper level, we may transcend time because our ego dissolves away completely, leaving us in a state of pure, unconditioned consciousness where there's no such thing as time.

And like meditation, mindfulness has long-term as well as immediate effects. If we practice it regularly it also helps to permanently quieten our thought-chatter, and therefore to subdue our ego and intensify our mental energy. Every time you practice mindfulness – or meditate – the stream of thought-chatter in your mind becomes a little weaker, and your ego becomes more subdued.

Slowness

To a large extent, doing things mindfully means doing them slowly. It's impossible to be mindful of anything while you're rushing, because your attention isn't on the present, but on the future that you're rushing towards. When you hurry, what you're actually hurrying *from* is the present moment. You don't want to face it, either because it's painful or because the future seems more interesting, and so you turn away from it and rush into the future. But of course, when you actually reach them, those future moments are painful or dreary too, and so you don't accept them either. You rush through them, continuing to focus on the future, and in this way, you never actually live in the present. And since our lives only consist of the present, in a sense you never really live.

The modern world puts a massive mount of pressure on us to hurry. For centuries our culture has been waging a constant war against time, trying to cut down on how long it takes to do everything – to travel from one place to another, to send messages, produce goods, cook meals, clean our houses, wash clothes, build houses, and so on. In some ways this has been beneficial – for example, there's no doubt that time-saving innovations like the washing machine and the vacuum cleaner have helped to free women from the drudgery of domestic duties and given them more time to devote to education and careers. But in recent years this speeding-up trend has spiralled out of control, and brought a massive amount of stress into our lives. We're so desperate to cook quickly that we don't take any pleasure from the creative process of cooking. We're so desperate to eat quickly that we don't take the time to enjoy our food. We're so desperate to live quickly that we forget to live. According to one study, in the 1960s 74 per cent of Americans said that they either sometimes or always felt rushed in their lives – but by 1995, this had risen to 87 per cent.[5]

The growing worldwide Slow Movement is all about resisting this pressure. This started with the Slow Food movement, which first came to public attention in the early 1990s, when it protested against the opening of a McDonald's restaurant in Rome. The Slow Food movement now has 83,000 members in 100 countries, and tries to persuade people to eat organic, locally produced food, and most importantly, to cook slowly (in order to enjoy the creative process of cooking) and to eat slowly (in order to fully savour meals). Slow Food has led to other movements too: Slow Travel, Slow Work, Slow Cities, Slow Education and Slow Books, to name just a few. Slow Travel usually means avoiding frenzied tourist holidays, and instead staying in one place – perhaps a rented cottage or farmhouse – and spending your time exploring the local area on foot or by bike. Slow Education means stepping out of the target-oriented treadmill of the normal school curriculum, and teaching children in a less demanding and more holistic and creative way.[6]

The main aim of these movements (and of the Slow Movement in general) is to encourage us to relish the activities that make up our lives, rather than trying to run through them as quickly as possible. The Slow Movement encourages people to stop wearing watches, to remove non-essential chores and activities from their lives, and to take up 'slow hobbies' like yoga, gardening or knitting. In other words, the movement is mainly about living mindfully, about turning your attention away from the future and back to the present.

And ironically, although the whole purpose of doing things quickly is – supposedly – to save time, in a way doing things slowly enables us to become much wealthier in terms of time. Far from wasting time by cleaning your house slowly or making a proper meal, if we do these things mindfully, living fully in the present, we actually *create* more time.

The third state of being

If you meditate regularly and try to live mindfully you'll soon begin to create more time in your life in another way. As I suggested earlier, most of us spend our lives swinging back and forth between two states of being: between states of absorption and states of boredom. In fact we spend the vast majority of our time in the first of these, spending our days absorbed in our jobs and our free time distracted by entertainments. We don't have much choice because, as we've seen, we find states of non-absorption (or boredom) so difficult to endure.

But one of the effects of meditation and mindfulness is to make it possible for us to live in a *third* state of being. We've seen that the reason why states of non-absorption are so unpleasant is partly because we have to face the psychic entropy of our chaotic (and usually negative-based) thought-chatter, and also because of the fundamental sense of isolation and aloneness which our strongly developed egos bring. But meditation *cures* us of these. The whirlwind of our thought-chatter slows down and fades away, until eventually we only use the thinking mechanism of our minds when we actually *want* to think, rather than thinking involuntarily all the time. And at the same time, as the boundaries of our ego become less sharp, we lose that terrifying sense of being trapped inside our heads with the rest of reality out there on the other side of our skulls.

And this means that we no longer need to fill our free time with distractions. The inner disharmony which drives us to spend so much time outside ourselves is removed. Normally we're like a teenager who spends all her time out because her parents argue all the time and there's a terrible atmosphere in the house. But meditation and mindfulness change the atmosphere, the discord in the house fades away, and the girl starts to feel comfortable at home again. In other words, we become able to live inside ourselves again, which also means living in the moment and in the world. After a period of making conscious efforts to live in the now, we start to become naturally and spontaneously mindful, without needing to be distracted from the moment. It becomes possible for us to just sit in our rooms without watching television or reading a book, to wait in a queue without getting impatient, or to do monotonous jobs without being overwhelmed by boredom and frustration. We no longer need to kill time, and as a result (although we'll obviously still experience states of absorption some-times, when we give our attention to things which interest us) we spend less time in states of passive absorption, and so experience more time in our lives.

Transcending linear time

Once we reach this point, and begin to live in a natural state of mindfulness, linear time no longer exists for us. This is inevitable, since – as we've seen – our sense of linear time is created by our thought-chatter, and in this third state of being there's no – or at least very little – thought-chatter. As a result the future and the past become much less impor-tant. We're still *aware* of them, and we can still recall past experiences and plan or consider the future when we want to, but we're no longer dominated by them. We know that they're really only abstractions, and that there's really only one tense: the present. We're still aware of deadlines and plans but we no longer feel pressurised by them. We no longer feel that time is running away from us, because in reality there's no time, but only *now*.

We also become free of the destructive power of linear time, the terrible awareness that nothing is permanent or stable, that 'time the destroyer' is continually eating away at our lives. Many romantic poets felt angst and despair when they contemplated time, moving them closer to death and meaning that beauty and happiness can only ever be temporary. The French poet Baudelaire complained of 'the horrible burden of Time weighing on your shoulders and crushing you to the ground',[7] while Keats lamented that in this world of time 'beauty cannot keep her lustrous eyes,/or new love pine at them beyond tomorrow'.[8] This view of time produces the 'time sickness' (in Larry Dossey's words) which is eating away at our culture, and which – as Dossey points out – isn't just bad for our psychological health, but is also responsible for some of our most deadly illnesses, stress-related conditions like cancer and heart disease.[9] But if you are completely rooted in the present, you become free of both this existential despair and this time sickness. In a sense, you are one with time, flowing with it, rather than outside it, trying to keep up. Instead of focusing on the future and rushing away from the present to get there, you become aware of a new kind of permanence. You begin to see the present as a permanent and continuous reality, and one's that glorious and meaningful, rather than as a series of moments which flash by and then disappear forever. Time is no longer a river, but an ocean. In the words of the Buddhist scholar Nyanaponika Thera: 'Right Mindfulness recovers for the man the lost pearl of his freedom, snatching it from the jaws of dragon time.'[10] The external conditions of your life may still be changing, but this doesn't make happiness impossible (as Schopenhauer believed), because you possess an *inner* happiness and stability which isn't affected by external circumstances

Progression not regression

This new relationship to time comes from bringing about a permanent change in our psyche: weakening the ego structure and taming its mad thought-chatter, so that there's always a surplus of energy inside us and we no longer have a strong sense of ego-separateness and duality.

But doesn't this – you might argue – just mean going back to being a child? After all, we develop our strong ego and the desensitising mechanism as we grow into adults, and so becoming free of them must be like being a child again.

It's true that there are parallels between this state and childhood. At this level, there's a similar kind of intensity of perception to childhood. Without the desensitising mechanism, the world is always an incredibly real and fascinating place and we never experience boredom. And like children, we're free of a lot of the psychological turmoil which most normal adults suffer from. With our ego's quietened, we don't experience the disorder of our normal thought-chatter, and the anxiety and worry that the negative tone of most of our thinking brings. In addition, now that our ego is no longer as strong and separate, we don't experience a fundamental sense of ego-isolation, of being trapped inside our heads with the rest of reality out there. Like children, we don't feel separate to the world, but as if we're a part of it.

The parallels are certainly there in terms of time too. As we move beyond our normal ego – or as we move towards a higher level of consciousness, you might say – we reverse the process of becoming more and more subject to time which occurs as we develop through childhood into adults. Back then, the speed of time progressively increased, but now it gets progressively slower. And it continues to get slower until you reach the highest, mystical levels of consciousness – where the ego disappears completely – and you re-experience the timeless realm which you experienced during the first months of your life.

But despite these parallels, it's definitely not a question of regressing to a childhood state. It's not a matter of 'undoing' or destroying our ego, but of controlling and transcending it – that is, going forward into a higher state rather than regressing to an earlier one. As the philosopher Ken Wilber points out, children live in a *pre*-rational and *pre*-egoic state, while the mystic lives in a *trans*-rational and a *trans*-egoic state.[11] The problem isn't the ego in itself, but that it becomes too strongly developed, too uncontrollable and chaotic, taking over our whole being and alienating us from the 'real self' – or witnessing consciousness – which lies beneath. But at the same time we *need* an ego. Although it has some serious drawbacks, the adult ego has some great benefits too. It gives us a new practical and organisational ability that children don't have, much greater powers of concentration and intellect and the ability to think abstractly and logically. Transcending our normal ego obviously doesn't mean sacrificing these benefits – it means keeping them but doing away with the negative effects. The ego is still there as a structure, which we can use whenever we need to think logically or be practical, like a coat which we can take on and off. But it no longer monopolises our psychic energy and dominates our whole being.

The end of time

To summarise, then, if you want to expand and transcend time through the internal way – as opposed to the external way of exposing yourself to as much unfamiliarity as possible – you need to:

- meditate regularly, twice a day if possible
- practise mindfulness by choosing two daily activities to do mindfully, e.g. having a shower or walking to the bus stop
- practise mindfulness in a more general way by making your life less hectic and stressful and trying to perform your daily activities slowly, never rushing unless absolutely necessary.

Your meditation practice and mindfulness exercises will certainly expand your sense of time temporarily – perhaps for a few hours after you've done them – but after a few months you should begin to feel a cumulative long-term effect too. As your normal state of being begins to change, you will develop a *permanently* expanded sense of time. This comes about for three reasons.

1. You generate a higher level of energy, which means that your desensitising mechanism no longer needs to function, and you absorb much more perceptual information from your surroundings and your experience.
2. Your ego become less strong and less dominant, which – since time is a creation of the ego – makes time slow down, meaning you're less bound by linear time.
3. You become able to live in the moment, so that you no longer need to escape into states of absorption in which time passes quickly.

It goes without saying that this internal way is free of the problems which the external way of living a life of unfamiliarity brings. There's no need to go anywhere, no need to keep moving to stay one step ahead of the desensitising mechanism, and no need to sacrifice any other human needs. You don't need unfamiliarity because the whole of your environment and all of your experiences – even though you might have been exposed to them thousands of times before – are still, and always will be, radiantly real and alive. Despite his incredible heroism, Rimbaud made a fundamental mistake, which meant that his attempts to find freedom were always going to fail: he tried to change his life, rather than his own psyche. Unfamiliarity is essential food for the mind, and always has a stimulating and time-stretching effect; but as a method of trying to overcome the desensitising mechanism, it doesn't work.

This is by far the best way of making our lives last longer – not by making them as safe and secure as possible, not by trying to delay the ageing process by exercising and eating healthily, not by having our bodies frozen when we die, and not even by filling our lives with new experience – but by expanding time from the *inside*, by changing the psyche through which we experience it.

And if you do change yourself in this way you will gain much more than just time. In fact these other gains will be so great that the desire to live for as long as possible will no longer seem so important. In a sense, the desire to live for as long as possible isn't so different from the egotistical desire to own as many cars or to earn as much money as possible. It means that you're trying to *possess* time, to add as many extra moments to your life as you can. It also means that you're alienated from the present, giving away your present moments to an unreal future which may never even come to pass. But in this new state of being the concept of 'how long' you live for won't matter. Instead of looking for fulfilment in future goals, you'll find it in the present. You'll gain your life back, the ability to *be*, to live in what Pascal called 'the only time which belongs to us' – to actually exist *in the world* rather than in an abstract world of thoughts or an unreal world of distractions. True well-being will also be yours, along with access to a new realm of meaning and beauty, a new sense of connection to the world and to a core of peace and stillness inside you.

And this may only be the beginning. Perhaps you will progress even further, to the highest states of consciousness, where your own sense of individual existence dissolves into oneness with the universe – at which point time will disappear completely, and you will experience eternity.

Acknowledgements

Most of the ideas in this book were expressed in a more tentative and truncated form in my first book *Out of Time*, published (in a small, now sold-out edition) by Paupers' Press in 2003. The managing editor of Paupers', Colin Stanley, was the first person to take an interest in my ideas on time, and I'm grateful for his talent-spotting abilities and his support. Colin also gave me an excellent description of an unusual experience of time (in Chapter 5), as did many other friends, acquaintances and students – all of whom I would like to thank. Thanks also to my audiences at talks and lectures and my students at the University of Manchester CCE, who participated in the informal experiments described in Chapters 1 and 2.

I did a lot of my research for the book at the University of Manchester's John Rylands library, in particular, using the electronic journals and inter-library loan service; thanks to the very helpful staff there.

My agent, Peter Tallack, has been a fantastic source of support and enthusiasm. At Icon Books, Simon Flynn and Duncan Heath (who also gave me a wonderful example of an unusual time experience in Chapter 5) both provided some extremely useful comments and suggestions. Thanks also to my eagle-eyed editor, Lucy Leonhardt, for spotting theoretical discrepancies and for tightening my occasionally loose prose. And also to my wife Pam – for her support and anecdotes and examples – and to my son Hugh, for reminding me of how fascinating and beautiful the world is.

Appendix I

The five laws of psychological time

1. Time speeds up as we get older.
2. Time slows down when we are exposed to new experiences and environments.
3. Time passes quickly in states of absorption.
4. Time passes slowly in states of non-absorption.
5. Time often passes slowly, or stops altogether, in situations where the 'conscious mind' or normal ego is in abeyance.

The two basic 'relativities' of psychological time

1. The speed of time is relative to the amount of information we absorb and process. The more information there is, the slower time passes.
2. The speed of time is relative to how strong and separate our ego is. The weaker the structure is (e.g. during early childhood, Zone experiences, higher states of consciousness), the slower time passes.

Appendix 2

Answers to the forward telescoping questions on p. 16

1. The Lockerbie air disaster   1988
2. The terrorist gas attack in a Tokyo tube station   1995
3. The fall of the Berlin Wall   1989
4. The Bill Clinton/Monica Lewinsky scandal   1997
5. The death of George Harrison of the Beatles   2001
6. The death of Princess Diana   1997

Notes

For full publication details of sources referred to, see the Bibliography on p. 241.

Introduction
[1] Schopenhauer, *Essays*, p. 56.

One. The First Four Laws of Psychological Time
[1] See Joubert, 'Subjective Accleration of Time'; Joubert, 'Structured Time and Subjective Accleration of Time'; Lemlich, 'Subjective Acceleration of Time with Aging'; Walker, 'Time Estimation and Total Subjective Time'.
[2] Carrasco et al., 'Time Estimation and Aging'; Espinosa-Fernandez et al., 'Age-related Changes and Gender Differences in Time Estimation'.
[3] Bryson, 'I Began to Suspect I Didn't Come from this Planet', p. 22.
[4] Wilber, *The Atman Project*, p. 7.
[5] See Gioscia, 'On Social Time' in Yaker et al. (eds.), *The Future of Time*; Piaget, *The Child's Conception of Time*.
[6] Quoted in Bryson, 'I Began to Suspect I Didn't Come from this Planet', p. 22.
[7] Gioscia, 'On Social Time' in Yaker et al. (eds.), *The Future of Time*, p. 79; Wearden, 'Origins and Development of Internal Clock Theories of Time'; Droit-Volet et al., 'Temporal Generalization in 3- to 8-year-old Children'.
[8] Fraisse, *The Psychology of Time*.
[9] Quoted in Dressler, *Time*, p. 3.
[10] Erikson, *Childhood and Society*.
[11] Quoted in James, *The Principles of Psychology*, Chapter XV.
[12] Kemp, 'Dating of Recent Historical Events'; Brown et al., 'The Subjective Dates of Natural Events in Very-long-term Memory'.
[13] Pascal, *Pensées*, p. 43.
[14] Kabat-Zinn, *Coming to Our Senses*, p. 162.
[15] Mann, *The Magic Mountain*, pp. 104–05.
[16] Angrilli et al., 'The Influence of Affective Factors on Time Perception', pp. 972–3.

[17] Quoted in ibid., p. 977.

[18] Csikszentmihalyi, *Flow – The Psychology of Happiness*, p. 58.

[19] Hall, *The Dance of Life*, p. 138.

[20] Murphy and White, *In the Zone*, p. 40.

[21] In his book *The Labyrinth of Time*, the philosopher Michael Lockwood puzzles over a similar question, noting the apparent paradox that:

> A period of an hour or so, which is brimming with events and activities that engage our interest, seems, in immediate retrospect, to have passed more quickly than usual. But a period of weeks or months that is likewise eventful, in a largely enjoyable manner, may seem, as we look back on it, to be longer than it really was. Thus, someone might say: 'I can't believe it's still only six months since we left Naples! So much has happened since!' (p. 375)

However, the difference here is that in the first example – the hour or so – Lockwood is referring to a state of absorption, in which our attention becomes narrowed to a particular object (or particular objects), while the second is an example of the second law of psychological time, in which you're exposed to new experience and time is stretched as a result.

Two. How Information Stretches Time: The First Two Laws Explained

[1] Becker, *The Denial of Death*, p. 50.

[2] Gopnik et al., *How Babies Think*, pp. 209–11.

[3] Becker, *The Denial of Death*, p. 50.

[4] Wordsworth, 'Intimations of Immortality', *Poems*, p. 71.

[5] James, *The Principles of Psychology*, Chapter XV.

[6] For example, Phipps, E.W.J., 'Bodytime' in Grant (ed.), *The Book of Time*.

[7] Ibid.

[8] Quoted in McCrone, 'When a Second Lasts Forever', p. 53.

[9] James, *The Principles of Psychology*, Chapter XV.

[10] Ibid.

[11] Ornstein, *On the Experience of Time*, p. 38.

[12] Ibid., p. 103.

[13] Stevens, 'The Myth of Rehabilitation'.

[14] Sahlins, *Stone Age Economics*, p. 36.

[15] Rudgley, *Secrets of the Stone Age*, p. 36.

[16] Gopnik, 'What I Believe but Cannot Prove'.

[17] The computer scientist Ray Kurzweil also connects our perception of time to how many new experiences there are in our lives. As he sees it, psychological time is determined by the number of noteworthy events (or 'milestones') we experience, together with how chaotic or disorderly our lives are. When we are young – especially when we're children – our lives are full of noteworthy events, and so time is massively expanded. But as we grow older there are fewer milestones for us to experience. The space between them gets progressively bigger, and our lives become gradually more orderly, full of familiar experiences which we repeat again and again. And with fewer milestones and less chaos in our lives, time gets progressively faster (in Kabat-Zinn, *Coming to Our Senses*).

[18] James, *The Principles of Psychology*, Chapter XV.

[19] Dawkins, *Unweaving the Rainbow*.

[20] Frankenhauser, *Estimation of Time*, p. 14.

[21] Csikszentmihalyi, *Flow – The Psychology of Happiness*

[22] James, *The Principles of Psychology*, Chapter XV.

[23] Gopnik, 'What I Believe but Cannot Prove.'

Three. Absorption and Time: The Third and Fourth Laws Explained

[1] Sacks, *Awakenings*, p. 101.

[2] Quoted in McKenna, *Food of the Gods*, p. 219.

[3] Fine, 'Organisational Time'.

[4] Draaisma, *Why Life Speeds Up As You Get Older*.

[5] See Csikszentmihalyi, *Flow – The Psychology of Happiness*.

[6] A similar theory is put forward by the psychologist Michael Flaherty in his study of time perception, *A Watched Pot*. Like James and Ornstein, Flaherty believes that time perception is linked to information processing. He believes that time passes slowly when 'stimulus complexity fills standard units of temporality with a *density of experience* that far surpasses their normal volume of sensations' (p. 95). He recognises that when we are waiting or bored it may seem like there's a smaller 'density of experience' than normal, but in actual fact there's a great deal of 'subjective activity' in our minds.

[7] James, *The Principles of Psychology*.

[8] Quoted in Flaherty, *A Watched Pot*, p. 45.

[9] Flaherty, *A Watched Pot*, p. 47.

[10] Dossey, *Space, Time and Medicine*, p. 46.

[11] Avni-Babad and Ritov, 'Routine and the Perception of Time', p. 543.

[12] Bayes et al., 'A Way to Screen for Suffering in Palliative Care'.

[13] Wyrick and Wyrick, 'Time Experience During Depression'.

[14] Klein et al., 'Smoking Abstinence Impairs Time Estimation Accuracy in Cigarette Smokers'.

[15] In his book, *About Time* – another very helpful and interesting study of time perception – the American psychologist William Friedman lists six 'time distortions' which are similar to the laws I have suggested. These are: 1) engaging tasks make time pass more quickly; 2) more events lengthen the perception of duration; 3) ageing accelerates the speed that time appears to pass; 4) a given interval seems longer if a judgement of the duration is anticipated; 5) a duration seems longer if we are frustrated, waiting for a positive experience, waiting for a specific event, or in fear of imminent danger; 6) an interval seems longer if it is remembered in more detailed pieces and shorter if we think of it more simply. No. 1 is very similar to my third law of psychological time ('Time passes quickly in states of absorption'), the only slight difference being that it doesn't include the states of passive absorption in which we don't engage in tasks, but in which time still passes quickly (e.g. watching television or playing computer games). If we think in terms of experience and also *perceptual* events, no. 2 has similarities with the second law (new experience), and also to Kurzweil's theory that during childhood and periods of adventure and unfamiliarity time is stretched by the greater amount of experience we undergo. No. 3 is, of course, the same as my first law, while no. 5 relates to the discussion of how time passes in states of non-absorption and discomfort. The no. 4 and 6 distortions are certainly true, but they aren't, I believe, a particularly significant part of our experience of time compared to the others.

[16] Norgate, *Beyond 9 to 5*.

Four. When Time Stands Still: The Fifth Law of Psychological Time

[1] Flaherty, *A Watched Pot*, p. 30.

[2] Adams, *Ansel Adams*, p. 7.

[3] Hall, *The Dance of Life*, pp. 135–6.

[4] In Levine, *A Geography of Time*, p. 33.

[5] Icke, *Gazza, Best and Collymore*.

[6] Quoted in Dossey, *Space, Time and Medicine*, p. 170.

[7] Murphy and White, *In the Zone*, p. 42.

[8] Craft, *Twitch of the Snooze Button: Time Perception and Cognition in Humans*, p. 3.

[9] Quoted in Williams, 'The 25 Hour Day'.

[10] Levy, 'Psychological Implications of Bilateral Asymmetry'.

[11] Lancaster, *Mind, Brain and Human Potential*, p. 84.

[12] Quoted in Flaherty, *A Watched Pot*, p. 59.

[13] Ibid., p. 70.

[14] Quoted in Murphy and White, *In the Zone*, p. 107.

[15] Ibid., p. 108.

[16] Levine, *The Geography of Time*.

[17] Huxley, *The Doors of Perception*, p. 27.

[18] Ouspensky, *A New Model of the Universe*, p. 316.

[19] Other drug-induced experiences of timelessness are described by R. Ward in his 1957 book *A Drug-Taker's Notes*, and by Thomas De Quincy in his famous *Confessions of an English Opium-Eater*. De Quincy writes that under the influence of opium: 'Sometimes I seemed to have lived for seventy or a hundred years in one night; nay, sometimes had feelings representative of a duration far beyond the limits of any human experience' (p. 314). Interestingly, this doesn't tally with the finding – which I mention later in this chapter – that depressants normally have a time-constricting effect. However, perhaps opium – at least in De Quincy's experience of it – is an exception in that it appears to give rise to intense visionary episodes, which contain a massive amount of perceptual and cognitive information.

[20] Quoted in Ornstein, *On the Experience of Time*, p. 46.

[21] Hoffer, 'Effects on Psychedelics of Time' in Yaker et al. (eds.), *The Future of Time*, p. 400.

[22] Shanon, 'Altered Temporality'. Another person who has experimented with ayahuasca is the novelist Henry Shukman, who also experienced a massive expansion of time. As he writes in 'Stirred and Shaken': 'In that grossly altered state, if something lasted for a minute it was eternity.'

[23] Newell, 'Chemical Modifiers of Time', in Yaker et al. (eds.), *The Future of Time*, pp. 378–81.

[24] Tinklenburg et al., 'Marijuana and Ethanol'; Frankenhauser, *Estimation of Time*; Church, 'Properties of the Internal Clock'; Friedman, *About Time*.

[25] Archer, *Male Violence*, p. 134.

[26] For example, Wilber, *One Taste*, pp. 276–80.

[27] Cutting and Dunne, 'Subjective Experience of Schizophrenia', p. 400; Epstein, 'Natural Healing Processes of the Mind', p. 318.

[28] Jaynes, *The Origin of Consciousness in the Breakdown of the Bicameral Mind*, p. 427.

[29] Quoted in Drury, *Shamanism*, p. 59.

[30] Davlos et al., 'Deficits in Auditory and Visual Temporal Processing in Schizophrenia'; El Melegi, 'Exploring Time in Mental Disorders' in Yaker et al. (eds.), *The Future of Time*, pp. 260–68.

[31] Jaynes, *The Origin of Consciousness in the Breakdown of the Bicameral Mind*, p. 421.

[32] El-Melegi, 'Exploring Time in Mental Disorders' in Yaker et al. (eds.), *The Future of Time*, pp. 260–68.

[33] Quoted in Friedman, *About Time*, p. 117.

[34] Udolf, *Handbook of Hypnosis for Professionals*, p. 160.

[35] Gustafson, *Hypnotherapy in Medicine*.

[36] Quoted in Udolf, *Handbook of Hypnosis for Professionals*, p. 161.

[37] Vitaliano, *Neural Networks, Brainwaves and Ionic Structure: A Biophysical Model for Altered States of Consciousness*. Vitaliano also asked his subjects to name objects and images on a computer screen. He found that in their normal state of consciousness they could only name images which stayed on the screen for 20 milliseconds or more. When they were in the hypnotised state, however, they could name images which appeared for only 5–15 milliseconds.

[38] Saunders, 'Time Distortion Exercise', pp. 1–2.

[39] Cooper and Erickson, *Time Distortion in Hypnosis*.

[40] Draaisma, *Why Life Speeds Up As You Get Older*.

[41] Peeters, *Autism: From Theoretical Understanding to Educational Intervention*; Wing, *The Autistic Spectrum*.

[42] Davis, 'Disorientation, Confusion, and the Symptoms of A.D.D.'

[43] Baron-Cohen, *The Essential Difference*.

[44] In Osmond, *The Reality of Dyslexia*.

Five. Time Across Cultures

[1] Quoted in Porter, 'The History of Time' in Grant (ed.), *The Book of Time*, p. 15.

[2] In Griffiths, 'Living Time', p. 58.

[3] Jorgensen and Kaiser, 'Calendars, Creation and the End of Time'; Cross, *Hinduism*, p. 42.

[4] Spinney, 'How Time Flies'.

[5] Hall, *The Dance of Life*, pp. 132–3.

[6] Ibid., p. 37.

[7] Hall, *The Silent Language*, p. 35.

[8] Quoted in Hall, *The Dance of Life*, p. 86.

[9] Lawlor, *Voices of the First Day*, p. 37.

[10] Quoted in Service, *Profiles in Ethnology*, pp. 257–8.

[11] Hallowell, 'Temporal Orientation in Western Civilization and in a Pre-literate Society'.

[12] Ohnuki-Tierney, 'Concepts of Time among the Ainu of the Northwest Coast of Sakhalin'.

[13] Griffiths, 'Living Time'.

[14] Ibid.

[15] Ibid., p. 55.

[16] Hallowell, 'Temporal Orientation in Western Civilization and in a Pre-literate Society'.

[17] Hall, *The Silent Language*.

[18] Honore, *In Praise of Slow*, p. 29.

[19] Hall, *The Silent Language*.

[20] Hall, *The Dance of Life*, p. 29.

[21] Bloch, 'The Past and the Future in the Present', p. 288.

[22] Service, *Profiles in Ethnology*, pp. 257–8.

[23] von Bredow, 'Brazil's Pirahã Tribe: Living without Numbers or Time'.

[24] Hallowell, 'Temporal Orientation in Western Civilization and in a Pre-literate Society', p. 669.

[25] See Wildman, 'Dreamtime Myth: History as Future'.

[26] Levy-Bruhl, *The Soul of the Primitive*.

[27] Silberbauer, 'Hunter Gatherers of the Central Kalahari', p. 131.

[28] Boydell, 'Philosophical Perception of Pacific Property – Land as a Communal Asset in Fiji', pp. 21–4.

[29] Ravuva, *Vaka I Taukei: the Fijian Way of Life*, p. 7.

[30] Quoted in Griffiths, 'Living Time', p. 64.

[31] Geertz, *The Interpretation of Culture*.

[32] Josephy, *The Indian Heritage of America*, p. 37.

[33] Atwood, *The Making of the Aborigines*.

[34] Wilber, *Up From Eden*, p. 66.

[35] Werner, *The Comparative Psychology of Mental Development*, p. 152.

[36] Diamond, *In Search of the Primitive*, p. 170.

[37] Kleinfeld, 'Learning Styles and Culture', p. 153.

Six. The Timeless Moment: Higher States of Consciousness and Time

[1] Hay and Heald, 'Religion is Good for You'.

[2] Greeley, *Ecstasy*.

[3] Quoted in Spencer, *Mysticism in World Religion*, p. 238.

[4] Anon., *The Life of Ramakrishna*, p. 20.

[5] Quoted in Fontana, *Psychology, Religion and Spirituality*, p. 124.

[6] Johnson, *Watcher on the Hills*, p. 56.

[7] Ibid., p. 86.

[8] Quoted in Spencer, *Mysticism in World Religion*, p. 238.

[9] Quoted in Happold, *Mysticism*, p. 279.

[10] Ibid., p. 390.

[11] Kant, *Observations on the Feeling of the Beautiful and Sublime*.

[12] Quoted in Alison et al. (eds.), *The Norton Anthology of Poetry*, p. 614.

[13] Wordsworth, *The Works of William Wordsworth*, p. 651, lines 130–2.

[14] Deikman, *Experimental Meditation*.

[15] Ibid.

Seven. The Illusion of Time

[1] Hawking, *A Brief History of Time*, pp. 182–3.

[2] Boslough, *Masters of Time*.

[3] Quoted in MacKenzie, *Adventures in Time*, p. 124.

[4] Davies, *Space and Time in the Modern Universe*, p. 221.

[5] Hawking, *A Brief History of Time*, p. 126.

[6] Quoted in Gebser, *The Invisible Origin*, pp. 12–13.

[7] Barbour, *The End of Time*.

[8] Quoted in Dossey, *Space, Time and Medicine*, p. 157.

[9] In Gebser, *The Invisible Origin*, p. 16.

[10] See Talbot, *The Holographic Universe*, pp. 205–07.

[11] Ibid.

[12] Sheldrake, *The Sense of Being Stared At*.

[13] See Wilson, *Mysteries*, pp. 147–9.

[14] Sheldrake, *The Sense of Being Stared At*.

[15] Ibid.

[16] Ibid., p. 239.

[17] Savva and French, 'An Investigation into Precognitive Dreaming'.

[18] Sheldrake, *The Sense of Being Stared At*, p. 226.

[19] Ibid.

[20] Chen et al., 'Non-invasive Detection of Hypoglycaemia using a Novel, Fully Biocompatible and Patient Friendly Alarm System'.

[21] MacKenzie, *Adventures in Time*.

[22] Ibid., pp. 7–8.

[23] Toynbee, *A Study in History*, pp. 129–30.

[24] Ibid., p. 139.

[25] Ibid., p. 131.

[26] Proust, *In Search of Lost Time*, p. 51.

[27] Lawrence, *The Complete Poems*, p. 660.

[28] Brown, 'The Reincarnation of James, the Submarine Man'.

[29] Tarazi, *Under the Inquisition*.

[30] Hapgood, *Voices of Spirits*.

[31] See Fenwick and Fenwick, *The Truth in the Light*, and Fontana, *Is There an Afterlife?*, for powerfully convincing summaries of the evidence.

[32] Fenwick and Fenwick, *The Truth in the Light*, p. 98.

[33] Bernstein, 'Physicist Uses NDEs to Clarify the Nature of Time', p. 5.

[34] Fenwick, *The Truth in the Light*, p. 72.

[35] Ibid., pp. 69–72.

[36] Ibid., p. 164.

[37] Quoted in Phipps, 'Bodytime' in Grant (ed.), *The Book of Time*, p. 150.

[38] Moody, *Life After Life*, p. 183.

[39] Smith, *Commentary to Kant's Critique of Pure Reason*.

[40] Dressler, *Time*.

[41] Davies, *Space and Time in the Modern Universe*, p. 3.

[42] See Lorimer, *Whole in One*, for a discussion of Bergson's theories.

Eight. Controlling and Expanding Time

[1] Quoted in Norgate, *Beyond 9 to 5*, p. 84.

[2] Cohen and Taylor, *Psychological Survival*, p. 181.

[3] Ibid.

[4] Ibid., p. 185.

[5] Ibid.

[6] Levine, *A Geography of Time*.

[7] Buetow, 'Patient Experience of Time Duration', p. 24.

[8] Dossey, *Space, Time and Medicine*, p. 54.

[9] Quoted in Draaisma, *Why Life Speeds Up As You Get Older*, p. 207.

[10] Quoted in Wilson, *Religion and the Rebel*, p. 69.

[11] Quoted in Shakespeare, *Bruce Chatwin*, p. 363.

[12] Ibid., p. 369.

[13] Draissma, p. 207. However, like William James and John Wearden, Draaisma believes that the speeding-up of time isn't a real phenomenon, but an illusion caused by memory. So when he tells us to

fill our lives with new experience, it's not actually so that we can live through more time (as I believe) but so that when we look back from the future, we will have the impression that we've lived through more time.

[14] Quoted in Robb, *Rimbaud*, p. 338.

Nine. Expanding and Transcending Time
[1] Andresen, 'Mediation Meets Behavioural Medicine'.
[2] Flanagan, 'The Colour of Happiness'.
[3] Davidson et al., 'Alterations in Brain and Immune Function Produced by Mindfulness Meditation'.
[4] Kabat-Zinn, *Coming to Our Senses*, p. 162.
[5] Norgate, *Beyond 9 to 5*, p. 12.
[6] Honore, *In Praise of Slow*.
[7] Quoted in O'Neill, *Collected Plays*, p. 1172.
[8] Keats, *A Pocket Book*, p. 40.
[9] Dossey, *Space, Time and Medicine*.
[10] Thera, *The Heart of Buddhist Meditation*, p. 41.
[11] Wilber, *Sex, Ecology and Spirituality*.

Bibliography

Adams, A., *Ansel Adams: An Autobiography*, Boston: Little, Brown, 1985

Alison, W.A., Salter, M.J., Stallworthy, J., Ferguson, M.W. (eds.), *The Norton Anthology of Poetry*, New York: W.W. Norton & Company, 1984

Andresen, J., 'Meditation Meets Behavioural Medicine', *The Journal of Consciousness Studies*, vol. 7 (11–12), pp. 17–73, 2000

Angrilli, A. et al., 'The Influence of Affective Factors on Time Perception', *Perception and Psychophysics*, vol. 59 (6), pp. 972–82, 1997

Anonymous, *The Life of Ramakrishna*, Madras: Ramakrishna Math, 1929

Archer, J. (ed.), *Male Violence*, London: Routledge, 1991

Atwood, B., *The Making of the Aborigines*, Sydney: Allen and Unwin, 1989

Avni-Babad, D. and Ritov, I., 'Routine and the Perception of Time', *Journal of Experimental Psychology*, vol. 132 (4), pp. 543–50, 2003

Barbour, J., *The End of Time*, London: Phoenix, 2000

Baron-Cohen, S., *The Essential Difference*, London: Allen Lane, 2003

Bayes, R. et al., 'A Way to Screen for Suffering in Palliative Care', *Journal of Palliative Care*, vol. 13 (2), pp. 22–6, 1997

Becker, E., *The Denial of Death*, New York: Free Press, 1973

Bentov, I., *Stalking the Wild Pendulum*, London: Wildwood House, 1978

Bernstein, P., 'Physicist Uses NDEs to Clarify the Nature of Time', *Vital Signs*, vol. 22 (2), pp. 3–12, 2003

Blake, A.E., *A Seminar on Time*, Charles Town: Claymont Communications, 1980

Bloch, M., 'The Past and the Future in the Present', *Man*, vol. 12, pp. 278–92, 1977

Boslough, J., *Masters of Time*, Reading, MA: Addison-Wesley, 1992

Boydell, S., 'Philosophical Perception of Pacific Property – Land as a Communal Asset in Fiji', *Pacific Rim Real Estate Society*, pp. 21–4, Jan 2001

Brown, N.R., Rips, L.J. and Shevell, S.K., 'The Subjective Dates of Events in Very-long-term Memory', *Cognitive Psychology*, vol. 17, pp. 139–77, 1985

Brown, R., 'The Reincarnation of James, the Submarine Man', *The Journal of Regression Therapy*, vol. 5 (1), pp. 62–71, 1991

Bryson, B., 'I Began to Suspect I Didn't Come from this Planet', *Guardian Weekend*, pp. 19–23, 2 September 2006

Buetow, S., 'Patient Experience of Time Duration: Strategies for "Slowing Time" and "Accelerating Time" in General Practices', *Journal of Evaluation in Clinical Practice*, vol. 10, pp. 21–5, 2004

Campbell, J., *The Masks of God*, New York: Viking, 1959–68

Capra, F., *The Tao of Physics* (2nd edn), New York: Bantam Books, 1989

Carrasco, M.C., Bernal, M.C. and Redolat, R., 'Time Estimation and Aging: A Comparison Between Young and Elderly Adults', *International Journal of Aging and Human Development*, vol. 52 (2), pp. 91–101, 2001

Chen, M., Daly, M., Natt, S. and Williams, G., 'Non-invasive Detection of Hypoglycaemia using a Novel, Fully Biocompatible and Patient Friendly Alarm System', *British Medical Journal*, vol. 321, pp. 1565–6, 2000

Church, R.M., 'Properties of the Internal Clock' in Gibbon, J. and Allen, L. (eds.), *Timing and Time Perception*, New York: New York Academy of Sciences, pp. 566–82, 1984

Cohen, S. and Taylor, L., *Psychological Survival: The Experience of Long-Term Imprisonment*, London: Penguin, 1981

Cohen, S. and Taylor, L., 'Time and the Long term Prisoner' in Hassard, J. (ed.), *The Sociology of Time*, London: Macmillan, pp. 178–87, 1990

Cooper, L.F. and Erickson, M.H., *Time Distortion in Hypnosis*, Baltimore: Williams and Wilkins, 1959

Craft B., *Twitch of the Snooze Button: Time Perception and Cognition in Humans*, 2000, downloaded from www.brock.craft.org/hci/twitch.htm, 17 January 2006

Cross, S., *Hinduism*, Shaftesbury: Element, 1994

Csikszentmihalyi, M., *Flow – The Psychology of Happiness*, London: Rider, 1992

Cutting, J. and Dunne, F., 'Subjective Experience of Schizophrenia', *Schizophrenia Bulletin*, vol. 11, pp. 397–408, 1989

Davidson, R.J., Kabat-Zinn, J., Schumacher, J. et al., 'Alterations in Brain and Immune Function Produced by Mindfulness Meditation', *Psychometric Medicine*, vol. 65, pp. 564–70, 2003

Davies, P., *Space and Time in the Modern Universe*, Cambridge: Cambridge University Press, 1977

Davies, P, *About Time*, London: Viking, 1995

Davis, R.D., 'Disorientation, Confusion, and the Symptoms of A.D.D.', *The Dyslexic Reader*, no. 11, Fall 1997, downloaded from www.dyslexia.com/library/add.htm, 11 April 2006

Davlos, D.B., Kisley, M.A. and Randal, G.R., 'Deficits in Auditory and Visual Temporal Processing in Schizophrenia', *Cognitive Neuropsychiatry*, vol. 7 (4), pp. 273–82, 2002

Dawkins, R. *Unweaving the Rainbow*, London: Penguin, 1999

De Quincy, T., *Confessions of an English Opium-Eater*, London: MacDonald, 1821/1956

Deikman, A., 'Deautomatization and the Mystic Experience' and 'Experimental Meditation', downloaded from www.deikman.com, 1 February 2003

DeMeo, J., *Saharasia. The 4000 BCE Origins of Child Abuse, Sex-Repression, Warfare and Social Violence in the Deserts of the Old World*, Ashland: OBRL, 1998

Diamond, S., *In Search of the Primitive*, New Brunswick: Transaction Books, 1974

Dossey, L., *Space, Time and Medicine*, Boston and London: Shambhala, 1982

Draaisma, D., *Why Life Speeds Up As You Get Older*, Cambridge: Cambridge University Press, 2004

Dressler, K., *Time: A Dimension of Consciousness or an Actual Reality?*, 1999, downloaded from www.scimednet.com, 13 February 2003

Droit-Volet, S., Clément, A. and Wearden, J.H., 'Temporal Generalization in 3- to 8-year-old Children', *Journal of Experimental Child Psychology*, vol. 80, pp. 271–88, 1999

Drury, N., *Shamanism*, Shaftesbury: Element, 1994

Dunne, J.W., *An Experiment with Time*, London: Faber and Faber, 1950

Eliade, M., *From Primitives to Zen*, London: Collins, 1967

Epstein, A., 'Natural Healing Processes of the Mind: 1. Acute Schizophrenic Disorganization', *Schizophrenia Bulletin*, vol. 5, pp. 313–20, 1979

Erikson, E., *Childhood and Society*, New York: W.W. Norton, 1950/1993

Espinosa-Fernandez, L., Miro, E., Cano, M. and Buela-Casal, G., 'Age-related Changes and Gender Differences in Time Estimation', *Acta Psychologica*, vol. 112 (3), pp. 221–32, 2003

Evans-Pritchard, E.E., *Nuer Religion*, London: Oxford University Press, 1967

Fenwick, P. and Fenwick, E., *The Truth in the Light*, London: Headline, 1996

Fine, G.A., 'Organisational Time: Temporal Demands and the Experience of Work in Restaurant Kitchens', *Social Forces*, vol. 69, pp. 95–114, 1990

Flaherty, M.G., *A Watched Pot: How We Experience Time*. New York and London: New York University Press, 1999

Flanagan, O., 'The Colour of Happiness', *New Scientist*, 24 May 2003

Fontana, D., *Psychology, Religion and Spirituality*, Oxford: Blackwell, 2003

Fontana, D., *Is There an Afterlife?*, Winchester: O Books, 2005

Fraisse, P., *The Psychology of Time*, London: Eyre and Spottiswoode, 1964

Frankenhauser, M., *Estimation of Time: An Experimental Study*, Stockholm: Almqvist & Wiksell, 1959

Friedman, W., *About Time*, Cambridge, MA and London: MIT Press, 1990

Gebser, J., *The Invisible Origin: Evolution as a Supplementary Process*, 1970, downloaded from www.unca.edu/-combs/integralage, 13 June 2004

Geertz, C., *The Interpretation of Culture*, New York: Basic Books, 1973

Gell, A., *The Anthropology of Time*, Oxford: Berg Publishers Ltd, 1992

Gopnik, A., 'What I Believe But Cannot Prove', *Guardian*, 7 January 2005

Gopnik, A., Meltzoff, A. and Kuhl, P., *How Babies Think*, London: Weidenfield & Nicholson, 1999

Grant, J. (ed.), *The Book of Time*, Newton Abbot: Westbridge Books, 1980

Greeley, A., *Ecstasy: A Way of Knowing*, Englewood Cliffs: Prentice-Hall, 1974

Greenfield, S., *The Human Brain*, London: Phoenix, 1997

Grey, M., *Return from Death*, London: Arkana, 1985

Gribbin, J., *In Search of Shrödinger's Cat*, London: Black Swan, 1997

Griffiths, J., *Pip Pip: A Sideways Look at Time*, London: Flamingo, 1999

Griffiths, J., 'Living Time' in Aldrich, T. (ed.), *About Time: Speed, Society, People and the Environment*, Sheffield: Greenleaf Publishing, pp. 53–67, 2005

Gustafson, P., 'Hypnotherapy in Medicine', downloaded from *Healthy Hypnosis*, www.healingwell.com, 13 May 2005

Hall, E.T., *The Silent Language*, New York: Doubleday, 1959

Hall, E.T., *The Dance of Life*, New York: Anchor Press, 1984

Hallowell, A.I., 'Temporal Orientation in Western Civilization and in a Pre-literate Society', *American Anthropologist, New Series*, vol. 39 (4), pp. 647–70, 1937

Hapgood, C.H., *Voices of Spirits: Through the Psychic Experiences of Elwood Babbit*, New York: Delacorte Press, 1975

Happold, F.C., *Mysticism*, London: Penguin, 1970

Hardy, A., *The Spiritual Nature of Man*, Oxford: Clarendon Press, 1979

Hartocullis, P., *Time and Timelessness*, New York: International Universities Press, 1983

Hawking, S., *A Brief History of Time*, London: Transworld, 1996

Hay, D. and Heald, G., 'Religion is Good for You', *New Society*, 17 April 1987

Heinberg, R., *Memories and Visions of Paradise*, Wellingborough: Aquarian Press, 1990

Honore, C., *In Praise of Slow*, London: Orion, 2005

Horgan, J., *The End of Science*, New York: Helix Books, 1996

Huxley, A., *The Doors of Perception and Heaven and Hell*, London: Penguin, 1988

Icke, D., 'Gazza, Best and Collymore: Right Brain Gods in a Left Brain World', downloaded from www.football365.co.uk, 3 July 2000

James, W., *The Principles of Psychology*, New York: Dover, 1950

Jaynes, J., *The Origin of Consciousness in the Breakdown of the Bicameral Mind*, London, Pelican, 1976

Johnson, R.C., *Watcher on the Hills*, Norwich: Pelegrin Trust, 1959/1988

Jorgensen, B. and Kaiser, M., 'Calendars, Creation and the End of Time' in Aldrich, T. (ed.), *About Time: Speed, Society, People and the Environment*, Sheffield: Greenleaf Publishing, pp. 50–52, 2005

Josephy Jr, A.M., *The Indian Heritage of America*, London: Pelican, 1975

Joubert, C.E., 'Subjective Acceleration of Time: Death Anxiety and Sex Differences', *Perceptual and Motor Skills*, vol. 57, pp. 49–50, 1983

Joubert, C.E., 'Structured Time and Subjective Acceleration of Time', *Perceptual and Motor Skills*, vol. 59, pp. 335–6, 1984

Kabat-Zinn, J., *Coming to Our Senses*, London: Piatkus, 2005

Kant, I., *Observations on the Feeling of the Beautiful and Sublime* (trans. Goldthwait, J.T.), Berkeley: University of California Press, 1961/2003

Keats, J., *A Pocket Book*, London: Grange Books, 1993

Kemp, S., 'Dating Recent and Historical Events', *Applied Cognitive Psychology*, vol. 2, pp. 181–8, 1988

Kenney, J.M., 'Logtime: The Subjective Scale of Life. The Logarithmic Time Perception Hypothesis', downloaded from www.ourworld.compuserve.com/homepages/jmkenney, 13 December 2002

Klein, L.C., Corwin, E.J. and Stine, M.M., 'Smoking Abstinence Impairs Time Estimation Accuracy in Cigarette Smokers', *Psychopharmacology Bulletin*, vol. 37 (1), pp. 90–95, 2003

Kleinfeld, J., 'Learning Styles and Culture' in Connor, W.J. and Malpass, R. (eds.), *Psychology and Culture*, Boston: Allyn and Bacon, pp. 151–9, 1994

Lancaster, B., *Mind, Brain and Human Potential*, Shaftesbury: Element, 1991

Lancaster, B., 'On the Stages of Perception: Towards a Synthesis of Cognitive Neuroscience and the Buddhist Abhidhamma Tradition', *Journal of Consciousness Studies*, vol. 4 (2), pp. 122–43, 1997

Lawlor, R., *Voices of the First Day*, Rochester: Inner Traditions, 1991

Lawrence, D.H., *The Complete Poems*, London: Penguin, 1990

Lemlich, R., 'Subjective Acceleration of Time with Aging', *Perceptual and Motor Skills*, vol. 41, pp. 235–8, 1975

Levine, R., *A Geography of Time*, New York: Basic Books, 1997

Levy, J., 'Psychological Implications of Bilateral Asymmetry' in Dimond, S.J. and Beaumont, J.G. (eds.), *Hemisphere Function of the Human Brain*, New York: John Wiley & Sons, 1974

Levy-Bruhl, L., *The Soul of the Primitive*, London: Allen & Unwin, 1965

Lockwood, M., *The Labyrinth of Time*, Oxford: Oxford University Press, 2005

Lorimer, D., *Whole in One*, London: Arkana, 1990

MacKenzie, A., *Adventures in Time*, London: Athlone Press, 1997

Malinowski, B., 'Time-reckoning in the Trobriands' in Hassard, J. (ed.), *The Sociology of Time*, London: Macmillan, pp. 203–18, 1990

Mann, T., *The Magic Mountain* (trans. Lowe-Porter, H.T.), New York: Alfred A. Knopf, 1968

McCrone, J., 'When a Second Lasts Forever', *New Scientist*, November 1997

McKenna, T., *Food of the Gods*, Bantam: New York, 1992

Minkowski, E., *Lived Time: Phenomenological and Psychopathological Studies* (trans. Metzel, N.), Evanston: Northwestern University Press, 1933/1970

Moody, R.A., *Life After Life*, New York: Bantam Books, 1977

Murphy, M., *The Future of the Body*, Los Angeles: Tarcher, 1992

Murphy, M. and White, R.A., *In the Zone: Transcendent Experience in Sports*, London: Penguin, 1995

Needleman, J., *Time and the Soul*, San Francisco: Berrett-Koehler, 2003

Norgate, S., *Beyond 9 to 5: Your Life in Time*, London: Phoenix, 2006

Novak, P., 'Buddhist Meditation and the Consciousness of Time', *Journal of Consciousness Studies*, vol. 3 (3), pp. 267–77, 1997

Ohnuki-Tierney, E., 'Concepts of Time among the Ainu of the Northwest Coast of Sakhalin', *American Anthropologist, New Series*, vol. 71 (3), pp. 488–92, 1969

O'Neill, E., *Collected Plays*, London: Jonathon Cape, 1969

Ornstein, R., *On the Experience of Time*, London: Penguin, 1969

Osmond, J., *The Reality of Dyslexia*, London: Cassell, 1993

Ouspensky, P.D., *A New Model of the Universe*, London: Arkana, 1984

Pascal, B., *Pensées*, London: Penguin, 1966

Peeters, T., *Autism: From Theoretical Understanding to Educational Intervention*, London: Whurr Press, 1997

Piaget, J., *The Child's Conception of Time*, London: Routledge and Kegan Paul, 1969

Proust, M., *In Search of Lost Time, Volume 1* (trans. Scott Moncrieff, C.K. and Kilmartin, T.), London: Chatto & Windus, 1992

Ravuva, A., *Vaka I Taukei: the Fijian Way of Life*, Java: Institute of Pacific Studies, University of the South Pacific, 1983

Ring, K., *Life After Life*, San Francisco: Harper, 2001

Robb, G., *Rimbaud*, London: Picador, 2000

Rudgley, R., *Secrets of the Stone Age*, London: Random House, 2000

Sacks, J., *Awakenings*, London: Picador, 1973/1990.

Sahlins, M., *Stone Age Economics*, New York: Aldine de Gruyter, 1972

Saunders, M.D., *Time Distortion Exercise*, downloaded from www.mind-course.com, 12 June 2002

Savva, L. and French, C.C., 'An Investigation into Precognitive Dreaming: David Mandell – the Man who Paints the Future?', paper presented at the International Conference for the Society for Psychical Research, 2003, downloaded from www.spr.ac.uk/confprogramme, 9 August 2006

Schopenhauer, A., *Essays*, London: Walter Scott Press, 1930

Service, E.R., *Profiles in Ethnology*, New York: Harper and Row, 1978

Shakespeare, N., *Bruce Chatwin*, London: Vintage, 2000

Shanon, B., 'Altered Temporality', *The Journal of Consciousness Studies*, vol. 8 (1), pp. 35–58, 2001

Shedfeld, P., *Reduced Environmental Stimulation*, New York: Wiley, 1980

Sheldrake, R., *The Sense of Being Stared At*, London: Arrow, 2004

Shukman, H., 'Stirred and Shaken', *Guardian Review*, 12 March 2005

Silberbauer, G., 'Hunter Gatherers of the Central Kalahari' in Harding, R. and Teleki, G. (eds.), *Omnivorous Primates: Gathering and Hunting in Human Evolution*, New York: Columbia University Press, pp. 455–98, 1981

Smith, N.K., *Commentary to Kant's Critique of Pure Reason*, New York: Palgrave Macmillan, 2003

Spencer, S., *Mysticism in World Religion*, London: Penguin, 1950

Spinney, L., 'How Time Flies', *Guardian*, p. 9, 24 February 2005

Stace, W., *Mysticism and Philosophy*, Macmillan: London, 1961

Stevens, J., 'The Myth of Rehabilitation', 1991, downloaded from www.csvr.org.za/papers/papsteve.htm, 20 November 2006

Talbot, M., *The Holographic Universe*, London: Harper Collins, 1991

Tarazi, L., *Under the Inquisition: An Experience Relived*, Charlottesville: Hampton Roads, 1997

Taylor, S., 'From the Unreal to the Real: Are Higher States of Consciousness Real?', *New Renaissance*, vol. 10 (1), pp. 12–14, 2000

Taylor, S., 'Rimbaud: the Existential Saint', *Abraxas 19*, 2002

Taylor, S., 'Spirituality: The Hidden Side of Sports', *New Renaissance*, vol. 11 (1), pp. 6–9, 2002

Taylor, S., *The Fall*, Winchester: O Books, 2005

Taylor, S., 'The Sources of Spiritual Experience', *International Journal of Transpersonal Studies*, vol. 23, pp. 48–60, 2006

Thera, N., *The Heart of Buddhist Meditation*, New York: Weiser, 1973

Tinklenburg, J.R., Roth, W.T. and Koppell, B.S., 'Marijuana and Ethanol: Differential Effects on Time Perception, Heart Rate and Subjective Response', *Psychopharmacology*, vol. 49, pp. 275–69, 1976

Tolle, E., *The Power of Now*, London: Hodder and Stoughton, 2001

Toynbee, A., *A Study in History*, vol. X, London: Oxford University Press, 1954

Udolf, R., *Handbook of Hypnosis for Professionals*, Northvale: Jason Aronson, 1995

Vitaliano, G., *Neural Networks, Brainwaves and Ionic Structure: A Biophysical Model for Altered States of Consciousness*, downloaded from www.vxm.com, 21 April 2002

von Bredow, R., 'Brazil's Pirahã Tribe: Living without Numbers or Time', 2006, downloaded from www.service.spiegel.de, 9 August 2006

Walker, J.L., 'Time Estimation and Total Subjective Time', *Perceptual and Motor Skills*, vol. 44, pp. 527–32, 1977

Ward, R.H., *A Drug-Taker's Notes*, London: Gollancz, 1957

Waterfield, R., *Hidden Depths: The Story of Hypnosis*, London: Macmillan, 2002

Wearden, J., 'The Wrong Tree: Time Perception and Time Experience in the Elderly' in Duncan, J., Phillips, L. and McLeod, P. (eds.), *Measuring the Mind: Speed, Age, and Control*, Oxford: Oxford University Press, pp. 137–58, 2004

Wearden, J., *Origins and Development of Internal Clock Theories of Time*, 1997, downloaded from www.cafescientifique.man.ac.uk/events/wearden.htm, 13 April 2006

Werner, H., *The Comparative Psychology of Mental Development*, New York: International Universities Press, 1957

Whyte, L.L., *The Next Development in Man*, New York: Mentor, 1950

Wilber, K., *The Atman Project*, Wheaton: Quest Books, 1980

Wilber, K., *Up From Eden*, Wheaton: Quest Books, 1981

Wilber, K., *Sex, Ecology, Spirituality*, Boston: Shambhala, 1995

Wilber, K., *One Taste*, Boston: Shambhala, 2000

Wildman, P., 'Dreamtime Myth: History as Future', *New Renaissance*, vol. 7 (1), pp. 16–19, 1996

Williams, C., 'The 25 Hour Day', *New Scientist*, 4 February 2006

Wilson, C., *Mysteries*, London: Granada, 1978

Wilson, C., *Religion and the Rebel*, Bath: Ashgrove Press, 1958/1984

Wing, L., *The Autistic Spectrum*, London: Constable, 1996

Wordsworth, W., *Poems*, London: Penguin, 1950

Wordsworth, W., *The Works of William Wordsworth*, Ware: The Wordsworth Poetry Library, 1994

Wright, R., *Stolen Continents*, Boston: Houghton Mifflin, 1992

Wyrick, R.A. and Wyrick, L.C., 'Time Experience During Depression', *Archives of General Psychiatry*, vol. 91, pp. 1441–3, 1977

Yaker, H., Osmond, H. and Cheek, F. (eds.), *The Future of Time*, London: The Hogarth Press, 1972

Index

Forty-fied
The Good, the Sad and the Bad of Fortysomething Life

Malcolm Burgess

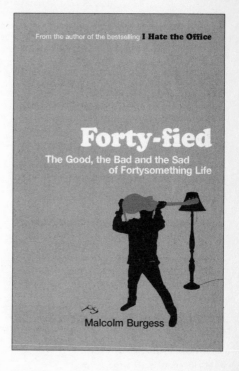

Is being forty the new thirty or are we all just kidding ourselves?

Malcolm Burgess presents a riotous A–Z of the realities of fortysomething life in the Noughties. Riotous, that is, like having your iPod on in the house. Today's fortysomethings have never had it so good – or so confusing. While our parents could look forward to a sensible middle age we're more likely to be playing our Morrissey records and thanking God Jonathan Ross is on Radio 2. There are so many different ways of being in our forties that many of us aren't quite sure where we're supposed to go next – or just how grumpy we're meant to be.

Forty-fied is the hilariously wry and observant essential guide to this complex decade in our lives. The *Metro* columnist and bestselling author of *I Hate the Office* leaves no embarrassing fortysomething scenario unturned – or do we mean unstoned?

For anyone forty and fabulous, or who's forty and owns ten fleeces, this is the laugh-out-loud funny book of your dreams … and no doubt your screams, too.

Hardback £9.99
Published in September 2007

ISBN 978 1840468 23 6

Crunch Time
How Everyday Life is Killing the Future

Adrian Monck and Mike Hanley

Crunch Time features two award-winning journalists arguing about the impact of our unthinking everyday actions on the future of our world.

Every age and every generation believes it's special, that it's on the cusp of something big. This time it's true – it's Crunch Time, and what we do now will make or break the future. The problem is that the things that we do every day – drive to work, buy toys for our kids, prepare our meals, have a cup of coffee – are conspiring to destroy it. Terrorism, poverty, ecological meltdown, climate change, pandemics – this is the background noise we have all learnt to live with. But what if all these things could be laid at our own feet? What if our civilisation is structurally, tragically flawed? What if we are using up tomorrow today?

Our society is moving faster than ever, yet it's also increasingly fragile and filled with risk. In *Crunch Time*, journalists Adrian Monck and Mike Hanley argue passionately with each other about the causes of these issues and what we can do about them. Believing that living in the 21st century means being answerable to the future, they help us to understand the critical decisions that we need to make now if we want to leave anything of value to future generations.

Hardback £9.99

ISBN 978 1840468 01 4